PERISCOPE DEPTH

PERISCOPE DEPTH

Submarines at War

Kenneth Poolman

WILLIAM KIMBER · LONDON

First published in 1981 by
WILLIAM KIMBER & CO. LIMITED
Godolphin House, 22a Queen Anne's Gate,
London, SW1H 9AE

ISBN 0-7183-0158-7

Typeset by Robcroft
and printed and bound in Great Britain by
The Garden City Press Limited,
Letchworth, Hertfordshire, SG6 1JS

Contents

List of Illustrations

CHAPTER ONE

Cranks

It was a calm, slightly misty morning off New York in early September 1776, an hour after first light. The pale early sunlight picked out the tents of the British army encampment on Staten Island, and the ships of the British fleet anchored offshore in the harbour rocking gently.

A redcoat sentry on the ramparts of the fort of Governor's Island looked drowsily out over the glassy water, giving the bay and the foreshore only a cursory inspection. He did not expect to see anything belligerent. The navy had complete control of the harbour, the Hudson and East rivers and all the waters round Manhattan Island and New York itself, where Washington's rebel rabble were trapped, driven out of Long Island a few days before after some bloody fighting. He was glad he'd missed that, he didn't fancy a ball in the belly from one of those Yankee long rifles.

Then something caught his eye, something moving on the water offshore. He screwed up his eyes. It wasn't a boat. What he could see looked more like the back of a fish, a big one. A great fool he'd look if he turned out the garrison for a whale. Whatever the thing was, it was zigzagging. The sun broke through the mist and shone on the object, which gleamed like metal . . .

The sentry shouted the alarm. Soon there were three or four hundred men along the parapet, watching the erratic manoeuverings of the alien floating body in the harbour. The naval party was roused, and spitting oaths six matelots manned a cutter and with a young lieutenant in the stern pulled raggedly towards whatever unnatural misbegotten creature had turned them out of their cosy hammocks.

The lieutenant was standing up with his spyglass trained on the object. He could make out what looked like a round hatch cover, and there appeared to be a number of stumpy spars . . . or pipes sticking up from the rounded upper surface. They were about fifty or sixty yards away when he saw something, a smaller rounded object, bob up in the water ahead of their target.

'Easy oars!'

It was some sort of infernal machine. He'd heard a buzz that the Yankees had a floating bomb or mine which they dropped in the tide to float down towards the fleet. He turned the cutter round and they made much better time back to the fort than they had coming out. They had landed and had the boat beached before the smaller floating object blew up with a gigantic explosion which threw up a huge column of white spray.

The incident was a talking point in mess and wardroom for at least twenty-four hours. No one realised that they had witnessed the end of the first submarine attack in history.

After the defeat of his left wing on Long Island and their lucky escape under the cover of bad weather across the East River, General Washington had followed the advice of his more pessimistic advisers and begun making plans for evacuating New York.

With Admiral Howe's British fleet filling the harbour, sailing with impunity up the Hudson River past the ineffectual guns of Forts Lee and Washington and surrounding the whole of Manhattan Island, this would be a dangerous operation. An American Continental Navy of only twenty-seven small converted merchantmen faced the Royal Navy's more than three hundred battle-hardened men o'war, and could not stop the British troopships from passing Sandy Hook, or the all-powerful wooden walls besieging Manhattan. Then someone thought of the crank Bushnell.

David Bushnell was a Yale College graduate in his mid-thirties. For many years he had been fascinated by the idea of a submersible boat, and had studied everything he could find on the subject: the stories of Alexander the Great's submersible glass barrel, Herodotus', Aristotle's and Pliny the Elder's accounts of attempts to build diving bells, and Englishman William Bourne's design of 1578 for a submersible boat with a wooden frame covered with waterproof leather and propelled by oars projecting through the sides. In 1620 Cornelius Drebble, a Dutchman living in London, built a Bourne boat and claimed an operational depth of twelve to fifteen feet, using goatskin bags which could be filled with water for submerging and squeezed out for rising. Thirty years later the Frenchman Le Son built an ambitious submarine craft seventy-two feet long, with paddle wheels amidships, but his clockwork engine could not move the boat.

Nathaniel Symons, a Devonshire carpenter, built a simpler

submarine in 1747 and demonstrated it in the River Dart, but it had no means of propulsion. Twenty-seven years later Suffolk wheelwright John Day constructed a watertight cabin inside the hull of a small boat, loaded her with ballast and remained on the bottom of Yarmouth harbour for twenty-four hours before jettisoning enough ballast to rise to the surface, but was later drowned in a bigger version.

After the first skirmish at Lexington between British troops and Massachusetts Militia had initiated open war between the former Colonists, now self-styled Americans, and Britain, Bushnell became absorbed by the difficulties of defence against Britain's overwhelming naval strength. Encouraged by Washington himself, he designed and built a one-man submarine which could be steered undetected for an enemy ship, fix an explosive charge to her hull, and retire before the charge went up and blew the ship out of the water.

The craft was named the *Turtle*, because someone said it looked like two turtle shells stuck together, but it looked more like a huge egg, floating big end up. It was built of wooden staves like a barrel, sheathed with copper to make it watertight, and bound round with several bands of iron. A round hatchway on top of the hull had six small round glass watertight scuttles let into the coaming, and 'in a clear day a person might see to read in three fathoms of water', as one of its operators said. On the bottom of the egg 700 pounds of lead were fixed for ballast, which made the *Turtle* quite stable in the water. The operator trimmed the craft's depth downwards by pushing on a foot pedal against a spring, letting water into the bottom. To adjust depth upwards there were two hand pumps to force water out. If the pumps choked, 200 lbs of the lead ballast could be jettisoned to bring the *Turtle* to the surface. The rise or fall of a cork in a glass tube indicated depth, an inch on the scale being equal to about one fathom, and there was a compass marked in phosphorus paint for reading in the dark. Three small vents three inches in diameter in the hatch top admitted fresh air on the surface; they were closed on diving, and there was enough air inside for thirty minutes' submerged operation.

Piercing the shell of the craft ahead and above were 'two oars of about twelve inches in length, and four or five in width, shaped like the arms of a windmill . . . ' operated by hand cranks, one to help submerge, one to provide forward motion. The 'navigator', as he was called, 'rowed' with one hand and operated 'a rudder having a crooked tiller' with the other.

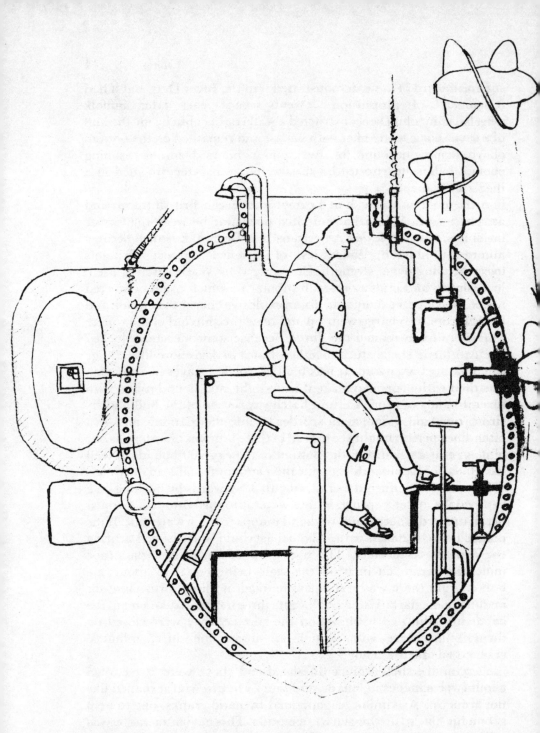

Sergeant Ezra Lee guides David Buschnell's *Turtle* towards her target HMS *Eagle*. From an early drawing.

Just forward of the hatch a long auger or drill protruded upwards, controlled by a cranked handle inside the hull. This was connected externally by a line to an egg-shaped 'magazine' made of hollow iron-bound oak covered with cork and taped over, attached aft just above the rudder and containing 150 lbs of gunpowder. The operator's aim was to steer for the target with the hatch just above the surface, then submerge under the ship, drive the auger upwards into the hull, unscrew it from inside, and unscrew the magazine attached to it. The release of the magazine started up a clock mechanism inside it, which ran for a set time – twenty or thirty minutes, to allow the operator to escape – before activating a gun lock which fired the powder charge.

Bushnell himself was not robust enough to take the *Turtle* into action, but his brother was trained to operate it, and was to have been the first submariner to go into action. Then he went down with fever and volunteers were called for. Bushnell selected a soldier, Sergeant Ezra Lee of a Connecticut regiment.

The *Turtle* was transported down to the New York waterfront, and Lee practised with his submarine boat, which he said looked 'most like a round clam'. He found the craft 'unmanageable in a swell or strong tide' and waited for conditions of slack water and calm. On a pitch-dark night, with the water still and glassy, whale boats towed him as near the British fleet, lying just above Staten Island, as they dared, then cast him off.

Trimmed with the hatch just above water level, the three small air vents in the hatch top open, Lee steered for the ships. But he had been cast off too soon, and the tide was sweeping him past them. He cranked his forward paddle, turned the *Turtle* about, rowed energetically into the slackening tide and managed to get himself under the stern of a looming sixty-four, HMS *Eagle*.

It was almost dawn by now. He could see men on the deck above and hear them talking. He closed off the three air vents and submerged the *Turtle* beneath the level of *Eagle*'s great hull. A few turns of the paddle and he was nestling under the ship. Turning the cranked handle of the auger, just below the hatch coaming in front of him, he felt the tip of the drill strike the ship's bottom. He went on turning, pressing hard, but could make no impression. The drill was not going in. And there was a harsh, metallic grating sound from above his head. Lee flogged the crank until his arm felt ready to fall off, using up his precious air fast, but the drill was just not biting. Whether he had struck an iron plate securing the rudder hinge, or whether the copper plating protecting the ship's hull from the teredo worm was resisting the auger's point was never established for certain, but Lee moved the *Turtle* along the ship's bottom to try another place.

The freshening tide caught him and swept the *Turtle* sideways. The little cockleshell went out of control and shot to the surface two or three feet from the ship's side. Lee had a blurred glimpse of an open port and the muzzle of a great black cannon above him, before *Turtle* plunged again like a porpoise. Frantically hands and feet flew between pumps and paddles, and he regained control. Turning about to make another attack, he realised that as soon as it was light, which would not be very long now, ships' boats would be plying in all directions, and he decided to break off and try to get back across the harbour without delay, as he had four miles of open water to traverse before he got past Governor's Island.

'So I jogg'd as fast as I could,' says Lee. His compass was now out of action, and he was forced to keep the *Turtle* on the surface so that he could check on his position, and make sure he was steering in the right direction. His greatest fear was that he would run aground on the island, as the fresh tide kept sweeping him off course. The *Turtle* zig-zagged as he continually corrected. It was at this point that he saw the boat pulling towards him. As they drew closer he released the magazine, 'in hopes that if they should take me they would likewise pick up the magazine, and then we should all be blown up together.' But the British tars took fright at the sight of the bobbing object and turned about. Lee weathered the island, and the *Turtle* shore party, watching anxiously for him, sighted him. A whale boat came out and towed him in.

Lee made one more attempt to sink a British warship. When Washington's men had evacuated New York and retreated up the Hudson to Fort Lee, a frigate came up and anchored off Bloomingdale. It was a temptation for the *Turtle*. Lee took her out and, trying a new tactic, made to row under the frigate's stern at surface trim with his vent doors open and screw the auger home close to the water's edge, just above the copper sheathing. But he was seen by the frigate's watch. Shutting his vents, he dived under her, but the cork in his depth gauge jammed, he went too deep, and had to abandon the operation.

Six years later America won her independence, without the further aid of the submersible, though Washington rightly described the *Turtle* project as 'an effort of genius'.

When Ezra Lee was making history in the *Turtle*, eleven-year old Robert Fulton was working on his father's farm in Connecticut. A gifted boy, he chose to study painting in England under his fellow American Benjamin West. But he had already shown a flare for mechanical engineering, as a precocious master gunsmith. Painting turned to draughtsmanship, drawing to engineering design, including a scheme for an advanced canal system. He also absorbed revolutionary social ideas from France, and discovered a hatred of Britain fuelled by remembered stories of her oppression of his native America and of his Irish forbears. The product of these factors combined was the design for a one-man submarine boat, which he offered to the French for the destruction of the blockading British Fleet. Fulton's bullet-shaped *Nautilus* was much more streamlined than the *Turtle* and incorporated several features of the modern submarine, including a conning tower, an aft-mounted propeller, and both vertical and horizontal rudders. But she still relied on muscle power for propulsion, with a collapsible sail to assist on the surface, and in operation could not get near the British frigates to fix her underwater mine – called a 'torpedo'.

Napoleon called him a charlatan, and his sponsors in the Ministry of Marine conveniently discovered that the submarine was an immoral weapon.

A furious Fulton swallowed his prejudice and offered his method, 'fit only for Algerians and pirates', to the British Navy. My Lords rejected *Nautilus* but were interested in his 'torpedo', and Fulton obligingly blew up an old brig for them. But the First Lord of the Admiralty, Earl St Vincent, refused to 'encourage a mode of war which they who commanded the seas did not want, and which, if successful, would deprive them of it.'

Fulton's hopes were finally crushed when Nelson's orthodox wooden walls routed the French and Spanish fleets at Trafalgar. Fulton returned home. There was no market for *Nautilus* in an America at peace, but he changed direction and designed and built the first successful steamboat.

With Fulton's little *Clermont*, steam power for ships had come to stay. The merchant fleets and navies of the world began to adopt it as an additional means of propulsion to wind power. Along with steam grew the use of iron for the protection and construction of warships.

In 1861 war began between the United States Federal Government and the Confederate States of America. Many skilled US Army officers joined the Confederate armies, and the

Confederacy won the first land battles. But very few US Navy officers joined the 'rebels', and the Confederacy, with an agrarian economy depending mainly on cotton, had no navy and no means of building one.

Like the rebel American colonies in their war against the British, the Confederacy adopted many irregular and unorthodox methods of fighting the war at sea. By dubious manipulation of international law, cruisers were built in foreign yards, mainly British, to raid Federal commerce on the high seas. A captured steam frigate, the *Merrimac*, cut down and converted into an ironclad warship, had immediate success against Federal sailing warships, but then met her match in the small Federal steam ironclad *Monitor*, the first warship to use a revolving turret.

The *Monitor* resembled a submarine in appearance, almost all her hull being under water, and while she and *Merrimac* were battling it out in Hampton Roads, the Confederate Navy was experimenting with submersible and partially submersible boats to try to break the Federal naval blockade which was inexorably strangling the South.

Down in Mobile, Alabama, Captain Horace L. Hunley had built and financed his third submarine, his first having been captured by the Federals at New Orleans, his second lost in Mobile Bay. The CSS *Hunley* could almost be called traditional in the sense that in design and basic principles she was comparable with Bushnell's and Fulton's submarines, with some indebtedness to the German Wilhelm Bauer's *Brandtaucher* (Fire Diver), which, operated by two treadmills, had given the powerful Danish fleet blockading German ports such a scare ten years before.

The *Hunley* was built of boiler plates. Oval in section, she had wedge-shaped fore and aft compartments containing ballast tanks filled by sea cocks and emptied by a hand pump. She was thirty feet in length overall. To power her through the water, a maximum of eight men turned a big crankshaft connected to a large propeller aft. There were two shallow conning towers with thick glass scuttles. A horizontal steering wheel operated a vertical stern rudder, a lever worked horizontal planes at the bow regulating diving and rising, and there was a metal keel in detachable sections to further assist ascent. There was a mercury gauge to indicate depth. Towing a copper torpedo, filled with ninety pounds of gunpowder, on a 200-foot line, she was to dive under an enemy vessel, then rise, so that the torpedo would hit the ship's bottom and explode.

After trials in Mobile Bay, which included the sinking of two

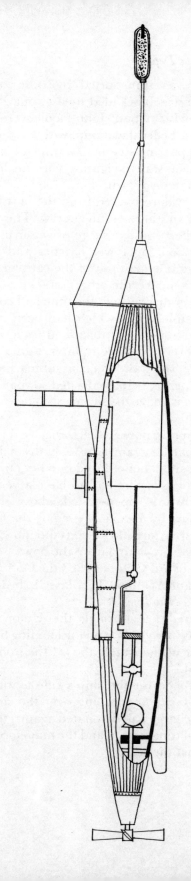

CSS *David*.

barges, *Hunley* was transported by train to Charleston, South Carolina, which was blockaded by a strong Federal fleet. Here six volunteers under Lieutenant John Payne were found. *Hunley* cast off for an attack on a Federal warship, with hatches and air vents open, the wash from a passing steamer swamped her and she sank, taking all but Payne, who was navigating with his head up in the for'ard conning tower hatchway, with her.

Again *Hunley* was recovered from the bottom, pumped dry, and Payne recruited another volunteer crew. They were preparing for a second sortie when again a surge of water from a steamer poured down the open conning tower hatches and into the boat. Payne managed to drag one man out of the capsizing hull. The rest were drowned, and Payne left the programme.

With the *Hunley* discredited, the time had come to turn to another type of submersible vessel which had been under development at Charleston. The hull of a gunboat had been cut down almost to the waterline, reinforced with iron plates, and a conning hatch fitted. Unlike the *Hunley* she was not a true submarine in that she could not completely submerge, being ballasted so that she operated with the conning hatch just above the surface, but her propulsion unit was far more sophisticated.

In place of muscle power, the *David*, as this craft was named, in anticipation of battles against the Goliaths of the Yankee Navy, used steam, having a small boiler fitted into her tapered hull. She too was dangerous to operate. She had to have a vent to the air, and this funnel stuck upwards, open-ended, above the surface of the sea, inviting swamping by a wave or even by the *David*'s own wash. Her weapon was a 'spar torpedo', a sixty-pound charge mounted on the end of a long spar projecting from the bows.

On the night of 3rd October 1863 the CSS *David*, commanded by young Lieutenant William T. Glassell, headed for the powerful Federal ironclad *New Ironsides*.

Glassell steered her through the Union Fleet without being spotted. She was fifty yards from the looming hulk of the *New Ironsides* when the officer of the watch aboard the ironclad sighted her, and raised the alarm.

The spar torpedo hit the ship's side as the Union sailors were rushing to the boats or jumping over the side. There was a fiery explosion. The charge had detonated against the waterline where the ship's armour plating began, and the main force of the explosion was expended against the sea.

A wave engulfed the *David*. Glassell, who had a life preserver, jumped overboard, swam away, and was taken prisoner. Engineer James Tomb relit his fire, which had been doused by sea water, picked up the crewmen who had jumped off the submersible, and took the leaking *David* back safely through the Yankee fleet, now in a great panic and furore.

Other *Davids* were built, but meanwhile the *Hunley* was put into service again. After another disastrous trial, Captain Hunley himself came up from Mobile and took charge. He hand-picked and trained a crew, they manned the crank and heaved her out into the harbour for trials. Captain Hunley closed the hatch, trimmed the boat, and dived out of sight. Hours went by, long past the time when her meagre supply of air would have given out. Eventually she was discovered stuck in the mud below the surface, with everyone dead. A ballast tank had given way.

After this tragedy it seemed like crass pig-headedness to persevere. But it seemed that as long as volunteers were forthcoming the 'unlucky' *Hunley* would try, try and try again. Another crew was found and on 15th October she made a dummy run on the Confederate ship *Indian Chief*. The torpedo towline fouled the ship's anchor chain, the *Hunley* plunged out of control and took another crew to their deaths on the bottom.

This time the *Hunley* was ordered to be broken up, but Lieutenant Dixon, who had the job of recovering her, persuaded the authorities to let him try yet again, with eight more volunteers.

Most of the Union ships had their anti-torpedo nets down, but the new sloop *Housatonic* had only just arrived on station and had so far neglected to do so. On 16th February 1864 *Hunley* nosed out into the harbour like a swimming alligator in the Everglades. This time she was equipped with a spar torpedo, like the *David*, and had orders not to dive.

The watch officer aboard the *Housatonic* yawned and stared out over the misty water. Then he saw a swirl in the sea.

'Bosun's mate, inform the Captain.'

Captain Charles Pickering rushed on deck and trained his telescope on the object. He distinctly saw a hatch cover.

'It's a Reb submersible! Slip the chain! Marines on deck!'

The ship's marines, sleepy and half-dressed, stumbled up on deck and opened up a ragged fire with small arms on the enemy craft, which had moved in too close for any of the sloop's main armament to be brought to bear. *Housatonic*'s anchor chain was rattling inboard, but it was much too late to escape.

The sweating, gasping men braced their legs wide on the deck as they flogged the crank round. With a splintering crash the spar hit the *Housatonic*'s wooden flank, biting deep, and the charge went off. Some of its force blew back and upset the *Hunley*'s trim. The submarine sank to the bottom, following the vitally damaged sloop as she settled.

The *Housatonic* sank so swiftly that only one of her boats was able to get clear, but the water at this point was shallow, and it was the doomed *Hunley* which suffered most. Both vessels hit the bottom, and the water rushing into the hole that the submarine had made in the *Housatonic*'s side sucked her own bow in with it. Unable to free themselves, Dixon and his crew died on the bottom alongside their victim. The *Housatonic*'s masts remained above water, so that the Union ship *Canandaigua* was able to come up and rescue nearly all her ship's company, who were clinging to her rigging, and only five men were lost.

CHAPTER TWO

Weapon of the Weaker Nation

The small boat rocked lazily on the gentle swell. Through half-closed eyes the fisherman watched his line, which sagged slackly from the bulwarks out to the almost motionless float. Behind him stretched the Manhattan skyline, ahead the grey-blue waters of New York harbour, calm as a village pond.

His eye caught a sudden flurry of foam on the water some two hundred yards beyond the float. Out of the sea burst a long, pointed missile, gleaming in the sun. With a rush of fear the fisherman saw its trajectory curve . The missile was heading straight for him. Then, just as suddenly, the trajectory decayed. The missile fell and plunged cleanly into the sea inches from his float, which bobbed in outrage.

As he stared, speechless, the fisherman saw another disturbance in the sea near the spot where the missile had emerged.

This was something bigger altogether, like a whale surfacing. The water heaved and frothed white. The object, whatever it was, rose out of the ocean, and he could see it was no whale, and no fish either. A low metal raised deck, about twelve or fifteen feet long, with three oblong vents or ports cut in its side, rode on the sloping back of a larger shape, its main bulk submerged, which *did* look like some metallic monster cetacean.

The phenomenon levelled off and slowly moved forward, white water cascading back over its curved flanks from the bow, and then turned in a wide arc towards the fisherman's boat. As it drew parallel, only a few yards away, its wash rocked first the agitated fishing float, then the boat herself. The phenomenon slowed and stopped.

Amidships in the flat upper deck a round hinged hatch cover swung up. From the hatchway emerged the head and shoulders of a man in a formal black suit, wing collar and tie, walrus moustache and granny spectacles on a long thin face, and a bowler hat.

The citified submariner raised his hat, then cupped his hands round his mouth.

'My torpedo. She will not run true.' The accent was Irish, with

New England Yankee overtones.

'You damn near sank me!' bawled the fisherman, shock turning to outrage.

'My apologies,' shouted Bowler Hat. 'I did not see you.'

'Then you need better eyesight!'

'That is the scientific truth, my dear sir. And we need a better torpedo. I have no control.'

It was 1882, twenty years after Glassell's steam *David* had damaged the *Ironsides* and Dixon had sunk the *Housatonic* with his primitive spar-torpedo, and here was Mr Holland, Irish immigrant turned New Jersey schoolteacher and then inventor, operating with confidence, after a pioneer experimental craft which he had scrapped, in a 31-foot long, 19-ton submarine boat powered by one of the new, small, economical, though unreliable, gasoline-fuelled internal combustion engines. A 9-inch pneumatic cannon fired a 6-foot long torpedo.

Holland's submarine was sound in principle. He knew it was the best working submarine so far developed and tested, better than the French *Plongeur* of 1863, with her huge hull packed with compressed air bottles to run her engine, her serious stability problems; handier than the steam-driven submersible gunboat being built by the Swede Thorsten Nordenfelt, inventor of an advanced quick-firing naval gun. *Holland II* was far more controllable underwater than all previous submarines because Holland operated her with a reserve of buoyancy, unlike the other pioneers, who filled their ballast tanks and reduced buoyancy to zero to submerge, then had constant difficulty adjusting ballast to maintain a steady depth. Instead of *sinking*, a Holland boat *dived*, using engine power and the action of horizontal rudders mounted aft. She was then levelled off and held at the desired depth by resetting the angle of the horizontal rudders.

But the *Nordenfelt I* was to be armed with the Whitehead torpedo, a weapon of inventive genius, which could carry a frighteningly destructive explosive head at a regulated depth on a steady course controlled by a gyroscope.

Holland's underwater missile did not match the superiority of his submarine boat. It would never sink a British warship. The Fenian Society, who had put up the money for *Holland II*, the *Fenian Ram* as she was popularly called, were getting impatient with his endless trials. They wanted action on their great scheme – a flotilla of *Hollands* transported in the watertight hold of an innocent merchant ship round the ports and naval dockyards of England, with the deadly submarines stealing out underwater and sinking the battleships of the British Navy, strongest weapon of the Imperial oppressors. The Fenians lost patience with Holland, and took over their submarine. But they could not work the controls, and ended up on a beach at New Haven, Connecticut. The Irish secret weapon never sailed.

The French also saw Britain as their potential enemy, with the mighty Royal Navy as her big stick, and the French Minister of Marine, Admiral Aube, like David Bushnell a century before him, saw the submarine as a potential means of reducing the overwhelming superiority of British numbers. The petrol engine was unreliable, and Aube's brilliant engineer Gustave Zedé chose the electric motor for his *Gymnote* and her bigger successor, the 150-foot *Gustave Zedé*, named after Zedé's death in an explosion on trials, and launched in 1893. The *Zedé* was an advanced design, but she had no horizontal rudders, and there were stability problems. Three sets of these rudders, which became known as 'hydroplanes', were added, one set for'ard, one amidships and one aft, and a deck 'casing' fitted to reduce the danger of swamping by a wave when surfaced in a choppy sea – as well as giving crewmen somewhere to stand. Even then she was hard to control, though she did 'torpedo' the battleship *Magenta* on Fleet manoeuvres in 1898.

This achievement advanced the cause of the submarine in the French Navy. A new boat, the *Morse*, was an improvement on the *Zedé*, but she also suffered from the shortcomings of electric power. The number of batteries which could be carried in a practicable submarine would only supply enough power for a short run before they needed re-charging, which could not be done at sea. Laubeuf's *Norval*, launched in 1898, reverted to steam for propulsion on the surface, with electric batteries for underwater use. Though the steam engine was more reliable than the contemporary petrol motor and provided good surface performance and range, it was still bulky, hot and dangerous to operate. But the *Norval* incorporated one big advance in submarine construction. She had a double hull. An inner, pressure-resistant shell housing crew and machinery was enclosed within an outer hull containing ballast and fuel tanks and shaped like a surface torpedo boat to give good sea-keeping characteristics. She could range five hundred miles on the surface and reach a comparatively fast eleven knots. Even under water she could run for several hours at five knots and increase to eight knots for an attack run.

John Holland, struggling against lack of money and the indifference of his own government, had reluctantly specified steam for his large 84-foot *Plunger* of 1893, but abruptly abandoned the design and built the smaller 50-foot *Holland*, with a 45 hp gasoline motor which could provide a range of 1,000 miles at seven knots on the surface and re-charge the electric batteries which drove her when submerged. She was agile, controllable and fired a Whitehead torpedo under water. The US Government, though not anticipating another war with the British or any other major power, reluctantly felt compelled to order a few Hollands.

The British, with their huge fleet of conventional warships, far outnumbering any other navy, had for a hundred years pretended that the submarine did not exist. When Admiral A.K. 'Tug' Wilson, in reply to demands for a British submarine force, called the submarine 'underhand, unfair, and damned un-English' his moral stance was transparently dishonest and a mere emotional re-phrasing of St Vincent's recorded refusal to 'encourage a mode of war which they who commanded the seas did not want, and which, if successful, would deprive them of it.' As late as 1900 Secretary of the Admiralty H.O. Arnold-Forster was saying: 'The Admiralty are not prepared to take any steps in regard to submarines, because this vessel is only the weapon of the weaker nation.'

But France, with a large fleet of conventional warships, had the largest submarine force in the world. Could not this be used against the Royal Navy to transform Britain into a 'weaker nation'? And in Imperial Germany, Admiral von Tirpitz was building a rapidly growing fleet of modern surface warships. The submarine lobby won. Loftily My Lords ordered 'five submarine vessels of the type invented by Mr Holland' – the one-time Fenian shipwright – as specimens which could be examined to see how this new 'underhand, unfair' enemy operated. The five Hollands were built in England by Vickers-Armstrong, under American supervision. By the time they had been completed British naval architects were sufficiently well-versed in the art to design a bigger and better submarine of their own, the 100-foot 180-ton A-class, which achieved 7 knots underwater by electric motor, with a 500 hp Wolsley petrol engine producing a surface speed of 11·5 knots, and a conning tower on the upper casing.

The type was very successful in service, and sired the improved 313-ton B and the similar C classes. All these early British boats were small single-screw 'coastal' types with petrol engines, and two torpedo tubes for'ard, but the D-class of 1906 embodied a big advance in design. The Ds were almost twice as large as the Bs and Cs, and their new, more efficient and reliable German-designed Diesel engines drove twin screws for a speed of 16 knots on the surface. An extra torpedo tube was added aft. The 800-ton E-class, the first of which was completed in 1913, was a high-water mark in submarine design, a fast ocean-going type strengthened by transverse bulk-heads, with additional beam torpedo tubes.

Tirpitz, obsessed with building up the German High Seas Fleet to outstrip the British Navy, had neglected the submarine. With his new powerful fleet of conventional warships he too regarded it as the 'weapon of the weaker nation' and hardly worth consideration except possibly for coastal defence – especially the small and puny specimens being produced at the turn of the century. Then, in 1904, Russia ordered three submarines of an advanced French ocean-going model from the German firm of Krupps. The power of these large boats to carry a war to an enemy battle fleet so impressed the Imperial Navy that the first *Unterseeboot*, the *U1*, was ordered from Krupps.

Like the British, the Germans, starting late, had avoided all the major mistakes of the pioneers, and produced an improved design. For propulsion on the surface the U-boat by-passed both the steam engine and the petrol motor, with its dangerously volatile and toxic fumes, and used, first, the more reliable and efficient heavy-oil kerosene engine; then it followed the British lead and took advantage of the new Diesel engine. *U1*, 139 feet long, displaced 283 tons. With twin screws, she could make 10·7 knots on the surface, 7 knots submerged. She was an experimental boat, and was equipped with only one torpedo tube, but following U-boats standardized on four tubes, two in the bow, two astern. The first eighteen boats used heavy-oil engines for surface propulsion, which tended to smoke and spark. From *U19*, completed in 1913, onwards, these were replaced by Diesel engines.

CHAPTER THREE

Live Bait

At three o'clock on the morning of 1st August 1914, the operational submarine force of the Imperial German Navy began leaving Heligoland for patrol duty in the North Sea, their bow-waves foaming white in the darkness as they glided out on the surface.

The U-boat arm, with a late start, mustered only twenty-five fully operational submarines, against the Royal Navy's seventy-four. But only the eight British D class and the nine E class so far in service compared in quality with the U-boats. No fewer than forty-nine of the British boats were of the obsolete or obsolescent petrol-engined B or C classes. The eight new German boats of the *U19* class, with better diesel engines, bigger (19·7-inch) and more reliable torpedoes than the British 18-inch type, and 1,000 miles greater range, outclassed even the excellent Es. Submarines now all used periscopes for sighting from below the surface, and the German Zeiss periscope lenses were the best in the world.

The U-boats returned from their patrol on 3rd August. At 11 p.m. on 4th August the wireless masts on the green-domed Admiralty in Whitehall broadcast the signal to all ships of the Royal Navy to commence hostilities with Germany – one hour before the ultimatum actually expired.

In the German Bight the British surface ships found some action, and sank the minelayer *Königin Luise*. It was not the scale of attack which the German Navy had expected, and at dawn on 6th August, ten U-boats put to sea on their first offensive patrol, to find the British Grand Fleet.

Heavy fog was the first enemy they encountered. Then Kapitänleutnant Otto Weddigen's *U9* broke down with engine trouble and had to turn back. The remainder carried on further and further north, looking for British battleships. It was not until they had reached the Orkneys on 8th August that the U-boats saw them, patrolling the Scotland-Norway gap.

Separated from the rest of the fleet, three modern Dreadnought battleships were carrying out a practice shoot with their big guns.

U15 was nearest to them. Her captain studied the battleships through his periscope, noted the ten turrets with the second superfiring over the first and the fourth over the fifth, the single tripod mast, the bridge built round the forward funnel. They were battleships of the *Orion* class, 22,500 tons, ten 13·5-inch guns. Jockeying into position, *U15* fired one torpedo. Lookouts aboard HMS *Monarch* saw the white track just miss them.

With destroyers and torpedo boats heading for her, there was no time for *U15* to make another attack. Above, a flurry of shock and anxiety went round the ships of the Grand Fleet. Obviously the range and potency of the U-boats had been underestimated. From then on, special anti-submarine lookouts were posted.

The U-boats headed for home. *U15* groped her way through thick fog and choppy seas. Then her engines broke down. As she lay hove-to she was sighted by sharp eyes aboard the scouting light cruiser HMS *Birmingham*, which rammed the U-boat and sank her with all hands. It was the first U-boat loss of the war, and it restored some of the lost confidence of the British Grand Fleet. *U13* was also missing, believed sunk by a mine.

But the U-boats had given the Admiralty a fright. They realized that the Grand Fleet's main anchorage at Scapa Flow in the Orkneys was vulnerable to submarines. The Fleet began to suffer from 'U-boatitis'. Every broomstick was a periscope.

Germany had lost two submarines, but the U-boats had flexed their muscles, and went back to look for targets off the British coasts.

In the afternoon of 5th September the watch officer of *U21*, one of the new diesel boats, raised the 'asparagus', as the U-boat sailors called their periscopes, and Kapitänleutnant Hersing took a swift look at the sea round the entrance to the Firth of Forth. Almost immediately he sighted a British cruiser off St Abb's Head, some way away but heading in his direction.

Hersing took *U21* closer to the cruiser's course. She grew bigger, filling the periscope's eye, steaming fast, her bow-wave curling crisply.

Hersing computed her course, speed and his firing angle. . . .

The watch officer's thumb pressed the firing button. There was the whoosh of compressed air, and the bows jerked upwards with the loss of weight. Quickly tanks were trimmed to bring the *Unterseeboot* back to an even keel.

Hersing caught intermittent glimpses of the target through the spray-lashed periscope lens. She was a light cruiser of the *Scout* class,

ten years old, built to work with destroyers, 3,000 tons, lightly armed
with ten 12-pounders. Her three high funnels, flat foc'sle and ram
bow made her either *Pathfinder* or *Patrol*.

Suddenly she seemed to shudder and heave. A column of fire and
flame shot skywards from just under her for'ard funnel, her bow-
wave began to die. It was a violent explosion, like a magazine going
up, a mortal hit. Inexorably the cruiser's stern began to tilt upwards,
her screws spinning in the air, until it stood upright in the sea. A
moment's hesitation, then she plunged to the bottom.

HMS *Pathfinder* was the first ship to be destroyed by a submarine
since the USS *Housatonic* had been sunk by the CSS *Hunley*, nearly
fifty years before. Only one of her boats floated off as she sank, taking
two-thirds of her crew of three hundred and sixty with her.

Meanwhile, thirteen of the seventeen submarines of the British
Harwich-based 8th Flotilla had been maintaining a defensive screen
east of the English Channel to protect the movement of the
Expeditionary Force to France, which began on 13th August,
against any German wolves coming down on the fold, with the old B
and C boats of the Dover Patrol giving close support to the
troopships.

With the Army safely across the Channel the Harwich D and E
boats resumed their patrols in the Heligoland Bight. It was
dangerous and frustrating work. Fog was prevalent, and these
waters were criss-crossed by German torpedo-boats and destroyers.

At quarter past seven on the morning of 13th September
Lieutenant-Commander Max Horton in *E9* raised his periscope's
eye above the surface of the Heligoland Bight. About two miles away
he saw a two-funnelled German light cruiser heading his way.

At 7.28 the submarine was six hundred yards from the starboard
beam of her target. Horton fired both his bow torpedoes, with an
interval of fifteen seconds. One minute later he heard a single loud
explosion. The coxswain tilted his hydroplanes, and *E9*, which was a
lively boat, plunged to seventy feet, maintained a course parallel to
that of the cruiser for three minutes, and then rose to twenty-two feet
for Horton to take a look at the damage.

Waves kept blurring his vision, but the target appeared to have
stopped with a list to starboard. He also saw shell splashes on their
port side and ahead of the cruiser. He turned the periscope to see
where the shots were coming from, but the periscope's eye was very
low in the choppy water, and all he could see was smoke, very close
by. Back he went to seventy feet. This time he stayed down for an

hour before coming up to twenty feet and raising the periscope. Where the cruiser had been there was a cluster of four or five trawlers, possibly picking up survivors. There was another trawler only about two hundred yards from *E9*, and there were destroyers about. Horton went back to seventy feet.

The ship which *E9* had sunk was the 2,040-ton *Hela*. She was an old ship, built in 1896, though reconstructed in 1910, but she was the British Submarine Service's first victim, and her sinking caused the German Navy to transfer all exercises from the North Sea to the safer waters of the Baltic. Almost in parallel, the British Admiralty, as a result of U-boat activity, was moving the Grand Fleet from Scapa Flow round to Loch Ewe in western Scotland, and withdrawing the blockade line to the gap between Scotland and Iceland.

Patrolling the southern end of the North Sea, watching for enemy minelayers and torpedo craft were the two destroyer flotillas belonging to the Southern Force, which was commanded by Rear-Admiral Arthur Christian, with the Force's five old obsolete 12,000-ton armoured cruisers of the 7th Cruiser Squadron, commonly known in the Fleet as the 'Live Bait Squadron', in support.

On 14th September very rough weather blew up in the North Sea. *E9*'s log recorded: 'Midnight. Very heavy seas. Bent stanchions and splash-plate. Endeavoured to rest on the bottom but disturbance continued to such a depth, i.e. 120 feet, that the submarine, despite 8 tons negative buoyancy, bumped.' The storm-tossed old cruisers of the Southern Force on patrol off the Dogger Bank were all suffering from minor defects in their engine-rooms, condenser troubles and the like, and Rear Admiral Christian got Admiralty permission to send in each ship in turn for repairs. HMS *Bacchante* left for Chatham Dockyard, with Rear Admiral Henry Campbell, commanding the 7th Cruiser Squadron, on board. The Southern Force cruisers also took it in turns to return for refuelling, and Christian's flagship *Euryalus* was urgently awaiting her turn, as well as being in need of repairs to her wireless aerial, damaged in the gales.

The weather continued very bad, and on 19th September the Admiralty ordered the destroyers to return to harbour, and the cruisers to switch to a patrol of the Broad Fourteens, an area twenty miles off the Dutch coast, to watch for enemy ships coming down to attack the transports in the Straits of Dover or lay mines in the Thames Estuary. This redistribution of forces originated with the Chief of the War Staff, Vice-Admiral Sir Doveton Sturdee, and was agreed with reluctance by the First Sea Lord, Admiral Prince Louis of Battenburg, who, like the First Lord, Winston Churchill, was uneasy about the use of the old cruisers for their risky front-line work in the North Sea. On the morning of Sunday 20th September Christian steamed off for Sheerness in *Euryalus*, the weather being too bad for him to transfer his flag to another ship, leaving the 'live bait' cruisers *Aboukir*, *Hogue* and *Cressy* labouring up and down on a NE-SW line, cramped in between the Dutch coast and a German minefield to the west, under the command of Captain John Drummond in *Aboukir*.

In the early morning of 22nd September Otto Weddigen raised the asparagus of *U9*.

Above them to seaward the three old 'live bait' cruisers of the Southern Force came on in line abreast, two miles apart, *Aboukir* in the centre, *Hogue* to starboard of her, *Cressy* to port, all alone on the sea, with no screen whatsoever. All ships had previously received a general warning about submarines, but no one in the cruisers had seen a trace of one for five weeks; the last warning had been a report of a U-boat off the Terschelling Light Vessel on the 16th. Anti-submarine measures had been relaxed. The ships were not zig-zagging, and speed had been reduced to 9 knots to conserve coal, which these old ships consumed so greedily. *Aboukir* had only two lookouts watching for U-boats, one at each end of the bridge, each one having to cover an arc of 90 degrees; *Hogue* had three special lookouts with Zeiss binoculars. In *Aboukir* one 12-pounder gun on each side was for submarine action. The 6-inch guns' crews on watch were asleep in their casemates. *Cressy* was better prepared, with one 3-pounder and one 12-pounder on each side for'ard manned and two 12-pounders on the after shelter deck.

Secret Admiralty information and British submarines' reports of enemy submarines in the area of the cruisers had not reached them, as they had been at sea for six days, and information of this sort was not sent by wireless. Thus the report of a submarine with a square conning tower and the figure 9 painted on her bows made by a Lowestoft trawler fifty miles east-sou'-east of Lowestoft on the 21st was not made available to Captain Drummond in *Aboukir*, senior officer of the squadron after the departure of Rear Admiral Christian and *Euryalus* to Chatham. Nor was any screening force either asked for by Drummond or ordered out by the Admiralty to protect the old ships, although the weather had moderated sufficiently by the 21st for the destroyers to put to sea. A flotilla left at dawn on the 22nd.

At 6.20 a.m. the destroyers were still a long way off. The tired lookouts in the cruisers, with too wide an area to search, did not see the tip of *U9*'s asparagus as it poked above the sea, only five hundred yards from *Aboukir*'s port bow.

'*Torpedo los!*' Weddigen sang out. First Officer Spiess pressed the firing key. They all waited tensely as the U-boat dived. Thirty seconds ticked by, then they heard the distant clang of a torpedo hitting armour, and cheering broke out.

Lying asleep on his bunk in *Aboukir* Captain Drummond felt the explosion, and left his cabin at once. Water still hung in the air, and smoke and dust were rising through the ship. He reached the bridge thinking that they had hit a mine. He ordered all the watertight doors closed. *Hogue* flashed, 'Have you been torpedoed?' On Drummond's order *Aboukir*'s Yeoman of Signals replied, 'Either torpedoed or struck a mine,' though Drummond's officers on the bridge told him that nothing had been seen of submarines or torpedoes. The Navigating Lieutenant had stopped both engines, which were now put to 'Slow ahead', and the ship was turned slightly so as to come head to wind and sea. The ship was listing and men coming up from below were told to fall in, while the order was given to clear away the main derrick to hoist out boats. *Hogue* and *Cressy* were closing the stricken *Aboukir*. Drummond signalled *Hogue* to keep ahead to avoid the line of mines presumed to be lying across their bows, followed immediately by 'Prepare to send boats.'

The ship listed to 20 degrees, then stopped. Drummond had a surge of hope that he could get her upright and keep her afloat by flooding the opposite wing compartments. Commander Sells reported that the steam pinnace, picket boat and launch could not be hoisted as no steam could be raised on the derrick. There was a shortage of boats anyway, some having been landed, others smashed by the gales, the port cutter shattered in the explosion, and this left only the starboard cutter, and a skiff between the funnels. The engineer commander reported that all steam had gone, and Drummond saw that the list was increasing again.

He said, 'I think she's going.'

'Yes, I think so,' said Commander Sells.

'Shall we let them go?'

'Yes.'

As *Aboukir*'s list increased rapidly, the order went to *Hogue* and *Cressy*: 'Send boats.' Loose wood and lifebelts were thrown overboard, and the men were told to jump.

In *Hogue* Commander Horton got two seaboats and a whaler

away, which pulled for *Aboukir*, now almost right over. Lieutenant Ashe RNR and twelve men got into the launch ready to be hoisted out. The sinking *Aboukir* got off a last signal: 'I have been torpedoed.'

The three boats which had left from *Hogue* and boats from *Cressy* were picking up *Aboukir* men.

U9 was just three hundred yards from *Hogue*. Spiess pressed the firing key for both bow torpedoes.

Lieutenant Ashe and his twelve men in *Hogue*'s launch had just cast off and were about to pull away when the ship was hit on the starboard side amidships. The column of water and smoke was still in the air when there was a second explosion in the same area. Swimmers from *Aboukir* felt the successive shock waves as blows in the stomach.

Up above in the tilting *Hogue* Captain Nicholson signalled *Cressy*, 'Look out for submarines.' Commander Horton was ordering the men to get hold of any detachable wood or lashed hammocks and get up into the picket boat or steam pinnace on the booms; then he went up and reported to the Captain, who told him to stop men from jumping over the side, and to go and find out from the Engineer Commander what the damage was. On his way Horton met Artificer Engineer Batty, who told him the water was over the engine-room gratings. Horton, on the port side of the ship, tried to get back to the bridge but the sea burst open the starboard entry port doors, and the ship heeled over too far for him to make it. Horton told the men in the port battery to jump, as the launch was still alongside. When he was abreast the port heads the ship lurched over heavily. He clung to a lifebelt on the rails but was washed off by a big wave and through the head doors. Trapped, he just managed to get his head and shoulders through a scuttle above him, and the sea washed him out. He climbed up the ship's side, and was washed off into the sea.

In just five minutes from being hit, the *Hogue* completely capsized, trapping many men below decks. Men stood on her bilge keel, others slid down the sides into the water. A man was seen clinging to one of her propeller blades, washed rhythmically by spray as the hulk wallowed in the swell.

About 6.45, twenty-five minutes after she had been hit, *Aboukir* took her final plunge, and about a quarter of an hour later *Hogue* disappeared.

Boats from the two sunken ships and from the surviving *Cressy* went round picking up men as fast as they could from the sea, but, without lifebelts, men were drowning all round them.

Interior of early submarine (possibly Holland's third boat, built at Delameter Iron Works, New York 1881). 1. Captain's seat; 2. steering wheel; 3. fore-planes; 4. aft planes; 5. depth gauge; 6. air bottles.

The launching of Royal Navy's first submarine Holland No. 1 at Vickers Son & Maxim Ltd, Barrow in Furness, 2nd November 1901.

(*Left*) Kapitänleutnant Walter Schwieger, captain of *U20* which sank the *Lusitania*, 7th May 1915. (*Right*) Kapitänleutnant Otto Weddigen, commander of *U9*.

(*Centre*) HMS *Cressy*, one of three 'Live Bait' cruisers sunk by Weddigen's *U21*, 22nd September 1914. (*Bottom*) RMS *Lusitania*.

The lucky ones found floating debris to cling to. Artificer Engineer Tom Rush, who had walked over *Hogue*'s side into the water, grabbed hold of an oar when he surfaced, but there was a marine already clinging to the other end, and the oar would clearly not support two, so Rush let go and swam until he found a small log of wood about four feet long.

A few minutes later a stoker, who was trying desperately to keep afloat, called out, 'Mr Rush, help me.'

Rush shouted, 'You had better get hold of this, it's all I can do for you.'

He let the stoker have the log, but found a floating chair and grabbed that. It wasn't very safe, then a hammock floated by. Rush caught hold of that and floated with the chair on one side of him and the hammock on the other. He made yet another switch to a large spar before he was picked up by a boat.

Commander Horton of *Aboukir* swam about amongst wreckage which was already overloaded with men but found a big baulk of timber which supported him until he was picked up by a cutter commanded by Coxswain Marks of *Hogue*.

Engineer Lieutenant-Commander Percy Huxham had left the silent, deserted engine room of *Aboukir*, watertight doors shut, oil lamps burning eerily, engines going very slow ahead but no steam in the gauges, no vacuum, water rinsing through the floor plates. Up on deck he had listened at the port ash shoots to see which stokeholds had water. He heard the water sloshing about in 3 and 4 but there was a heavy silence from 1 and 2, which must have been full. He could hardly walk for the heel on the ship. The port quarterdeck rails were under water. Huxham clambered aft along the tilting deck and went in off the top of the Admiral's stern walk. He swam away as quickly as possible, felt the shock waves of the torpedoes hitting *Hogue*, and was picked up by *Cressy*'s cutter. As he was being taken on board *Cressy* her 12-pounders were firing over his head.

A keen-eyed lookout had sighted a periscope 1,000 yards off, and *Cressy* had opened fire.

The rescued Percy Huxham met *Cressy*'s Engineer Commander Glazebrook, who told him to go to his cabin and put on anything he could find there.

Meanwhile *U9*'s two stern torpedoes were running. Their wakes were seen aboard *Cressy* and her captain ordered, 'Full speed ahead!' to try to avoid them. Slowly the old ship gathered way.

Huxham found Glazebrook's cabin, and was just taking a vest out of a drawer when there was a tremendous explosion, lifting him three or four inches off the deck. *U9*'s first stern torpedo had hit the ship in the starboard after boiler room. Percy grabbed a blanket from Glazebrook's bunk, wrapped it round himself, and went out into the passage, where he saw water splashing down the Gun Room Flat hatch.

Turning 190 degrees *U9* fired her last torpedo from a bow tube.

Percy Huxham sat shivering in his blanket on a fairlead on the port side of the quarterdeck. 'Periscope!' someone shouted. He looked across the deck to starboard and saw it, a grey stick three hundred yards away. Then there was a loud explosion on the starboard side as the torpedo tore into *Cressy*'s engine room.

Columns of smoke and white water shot upwards from the cruiser's side, and a cloud of smoke and soot erupted from the fourth funnel. *Cressy*'s list increased violently. Men climbed like ants over her side, then she turned right over, and sank. It was one hour since *U9*'s first torpedo had hit *Aboukir*. The 'Live Bait Squadron' was dead, and with it died 1,459 of the 2,200 men on board the three cruisers.

The small miscellany of boats surviving the ships carried on rescuing as many of the living as they could. Coxswain Mark's cutter from *Hogue* pulled about for some hours picking up men and transferring them to the ship's picket boat and steam pinnace, which had floated off when *Hogue* turned over. He rescued Engineer Commander Stokes, whose legs were broken, Fleet Paymaster Eldred, Commander Sells, and about 120 others. Captain Nicholson in *Hogue*'s other cutter saved many men from the sea. Able Seaman Farmstone of *Hogue* jumped overboard from the launch to make room for others and would not be helped until every man in their vicinity had been picked up. Lieutenant Tillard, himself picked up by a launch, got up a cutter's crew and saved many lives, as did Midshipman Cazalet in *Cressy*'s gig, who found Percy Huxham clinging to a fog buoy. He was not the only man to have two ships sunk under him in one hour. Among the men waiting to be rescued, Lieutenant Harrison of *Cressy* recalled, 'There was no panic, everyone was very quiet. When exhausted they said good-bye and went under.' Twelve *Cressy* men, including Captain Johnson, clung to a griping spar. 'The Captain,' reflected Yeoman of Signals Sydney Webb, 'was giving us instructions to keep our mouths closed and to breathe through the nose and not to expose too much of our bodies above the water, but just our heads. He also mentioned he had made a signal for the destroyers to come out and they would probably be out by ten o'clock. Soon after that I heard him, like, choking and shortly after he floated away.' The Harwich destroyers which had left at dawn did arrive and combined with the two Dutch steamers *Flora* and *Titan* in taking aboard men from floating wreckage, from the rescue boats, and finally the exhausted boats' crews themselves.

Weddigen was awarded the Iron Cross 1st and 2nd Class, and every member of his crew the Iron Cross 2nd Class. They were toasted in the mess, fêted in Kiel and Berlin. The British could blame their man in command for neglecting to zig-zag, for steaming too slowly (though the coal-gobbling problem was genuine), for failing to demand a destroyer escort when his superiors had failed to provide one; they could beach Drummond, Christian and Campbell on half-pay; they could belittle the loss of the old, obsolete ships: but the truth remained that even the mighty Royal Navy could not really write off the destruction of three still useful major units by one obsolescent submarine in one swift action. Most worrying was the thought that, given the score of more modern U-boats, with many more building, even with all the strategic and tactical errors corrected, a similar disaster could conceivably happen again, helping to alter the already uncertain balance of weaker and stronger nation. And this was not to mention the blow to prestige, the red faces in the Admiralty, especially that of Sturdee, the Chief of the War Staff, their Admiral Blimp, who, challenged by Commodore Roger Keyes, commander of the Submarine Service, on his disposition of the three cruisers, had said, 'My dear fellow, you don't know your history. We've always maintained a squadron on the Broad Fourteens.' The First Lord had boasted that if the German Navy did not come out to fight, 'it would be dug out like rats from a hole.' It was a typical Churchillian broadside, but after the sinking of the Live Bait Squadron, King George, himself an experienced naval officer, remarked drily that 'the rats came out of their own accord and to our cost.'

The *U9* men enjoyed their heroes' leave, and then went to sea again in company with *U17* to patrol the Pentland Firth between Scotland and the Orkneys, and the approaches to Scapa Flow. The early days of October were full of frustrated attacks, alarms, crash-dives and escapes.

Early on 15th October they were lying submerged at periscope depth east of Aberdeen, when the watch officer at the periscope sighted three British cruisers, zigzagging fast, converging on a point near where *U9* lay, exchanging signals. Weddigen estimated their rendezvous point and steered at full speed for the spot.

One cruiser just missed their periscope as she thundered overhead, then suddenly turned and slowed down, lowering a cutter to take mail or fresh orders to the other two cruisers, and Weddigen fired.

As usual the wait seemed endless. Then came the muffled clang of the explosion. Weddigen came up for a look through the asparagus.

Near the spot where their target had been was a small six-man cutter, all that was left of the 7,350-ton heavy cruiser *Hawke* of the 10th Cruiser Squadron, an old but powerful ship, with two 9·2-inch and ten 6-inch guns. The boat officer at the rudder hoisted the distress signal on the cutter's staff, but the other two cruisers were some miles away, and *Hawke* had sunk so quickly that there had been no time to get off a distress signal. Of *Hawke*'s complement of over five hundred, only forty-nine survived.

CHAPTER FOUR

Mutton Killer

Kapitänleutnant Feldkirchner in *U17*, which had been close at hand when *U9* had sunk the *Hawke*, and had only just been beaten to the attack, was returning from patrol empty-handed, when fourteen miles off the Norwegian coast he intercepted the small British steamer *Glitra*, sailing from Grangemouth in Scotland to Stavanger with coal, coke, iron plates and oil.

Germany had no plans to use her U-boats against commerce, but Feldkirchner decided he was justified in sinking the *Glitra*. International 'Cruiser Rules' stipulated that a merchant ship must not be sunk until the safety of her crew and any passengers she might carry had been properly seen to, and Feldkirchner obeyed both Cruiser Rules and German Navy law to the letter. He gave *Glitra*'s captain ten minutes to abandon ship, and only then did he order his boarding party to open the sea cocks and send her to the bottom. Though the sea was calm, he even towed the ship's boats for several miles towards the coast. Then he continued his passage home, worried that he would be court-martialled for exceeding his orders.

No action was taken against him, though the German naval C-in-C, von Ingenohl, considered the use of submarines against merchantmen barbaric, and the Chief of Naval Staff, von Pohl, thought it would be an infringement of international law, but a week after the *Glitra* sinking, Kapitänleutnant Schneider in *U24* torpedoed and sank the French steamer *Amiral Ganteaume*, carrying some 2,500 Belgian refugees, *without warning*, off Cape Gris Nez. Forty lives were lost and the action called an atrocity by the Allies. Schneider alleged that he had mistaken the *Ganteaume* for a troopship.

On 2nd November the British Government declared the North Sea a military area, forcing all neutral traffic for Holland, Denmark, Norway and the Baltic to pass first through the Straits of Dover and up the English coast to the Farne Islands off Northumberland. Any neutral ships trying to negotiate the North Sea through any other area would be in danger 'from mines it has been necessary to lay . . '

Angry German naval chiefs began to lose their finer feelings.

On 7th November von Pohl proposed to Chancellor Bethmann-

Hollweg a counter-blockade of Britain by submarine. The Chancellor was worried about the effects upon neutral countries, whose ships would be bound to suffer. Gross-Admiral von Tirpitz, Secretary of State of the Navy Department and creator of the German Navy, told an American journalist that Britain could be swiftly starved into surrender by U-boats, and the German public, which was beginning to go short of food as a result of the British blockade, demanded a ruthless submarine war against their oppressor.

The Naval Staff were confident that this could be achieved with the twenty sea-going U-boats at that time on strength, although an estimate prepared within the submarine service before the war had called for *two hundred and twenty* for this purpose. No more than about four boats could actually be kept on station at any one time. To cripple the greatest seafaring nation on earth Korvettenkapitän Bauer, commander of U-boats, wanted at least seven.

The Kaiser hesitated to commit Germany to unrestricted submarine warfare, just as caution made him limit the targets for his new Zeppelin terror weapons. U-boats continued to sink merchant ships, but in small numbers and usually obeying the restrictions of the Cruiser Rules and international law. A handful of submarines continued to inhibit the British Grand Fleet. Von Hennig in *U18*, the last of the 'Smokey Joe' kerosene-engined boats to be built, got into Scapa Flow on the flood tide; but she was rammed by a trawler, then by the destroyer *Garry*, the Pentland currents caught her and threw her on to the rocks, where Hennig abandoned her.

Hersing in *U21*, sinker of *Pathfinder*, was cruising in the English Channel. With his usual nerve, he was running surfaced through a thick November mist, the boat pitching and rolling in heavy seas and taking spray over the conning tower when a steamer loomed up through the fog.

'Man the gun!'

The for'ard gun's crew tumbled out of the conning tower and ran forward over the slippery casing to the 3·4-inch.

'Fire a shot across her bows.'

There was a crack, the whistle of a shell and a white plume of water dangerously close to the merchantman's bows. She hove to, rolling in the trough.

The sea was too high, and enemy warships too close and too numerous in these waters for him to board.

He cupped his hands and shouted up at the anxious faces peering down from the wing of the ship's bridge.

'Lower a boat and bring me your manifest!'

The German sailors grinned at one another. They knew their skipper by now, the cool nerve that enabled him to stop here in the English Channel, thick with British destroyers, and calmly check a ship's papers as if they were on the high seas.

A boat came slowly and jerkily down from the ship's davits and struggled across the heaving water to the submarine's side, where the sailors fended off with their oars as the boat bumped and scraped on the U-boat's saddle tank. With difficulty the coxswain handed the ship's papers to a crouching German sailor, who took them up to Hersing on the bridge. They showed clearly that the French ship *Malachite* was carrying war material from Liverpool to Le Havre.

'Abandon ship!' Hersing shouted to the French skipper. As the shivering matelots rowed ashore in their lifeboat, *U21* sank the *Malachite* with her gun. Three days later, still pushing his luck in the same waters, Hersing sank the SS *Primo*, carrying coal from England to Rouen.

Admiral von Pohl was keeping up pressure for a no-holds-barred *Unterseeboote* blockade, reassuring dubious Bethmann-Hollweg that his captains would be meticulous in differentiating between Allied and neutral ships.

The Chancellor did not want any more *Amiral Ganteaumes*. On 31st December Schneider, the destroyer of the refugee ship, was once again hunting in the English Channel. He had no trouble getting through the minefield off the Straits of Dover. In fact the defences here were giving the British a false feeling of security. The mines did not give full coverage, and some of them had already been hit by vessels and failed to go off. Exercising all on their own, with no destroyer screen at all, as if the *Cressys'* disaster had never happened, were eight pre-*Dreadnought* battleships. In the darkness of 2.25 a.m. on New Year's Day 1915 he aimed, and. . . .

'*Torpedo los!*'

They all heard the distant metallic crash. Schneider took the boat up and saw his target dropping out of line, listing badly.

Rather than trying for another battleship, Schneider, who was no Weddigen, took the soft option and hung around the stricken goliath, to finish her off if she needed it. Three-quarters of an hour after the first hit, he fired a second tinfish into her exposed, barnacled bottom as she lay heeled over. Water rushed into the hole and she rolled back on to an even keel. But she was clearly finished, and Schneider left her alone. HMS *Formidable* sank just before five o'clock. Boats from the other ships, handicapped by the darkness and rough seas, rescued 223 of the complement of 780.

Otto Hersing was home for Christmas and New Year's Day, following his successful cruise in the English Channel, making sardonic jokes with colleagues about the U-boat arm having to win the war for the Kaiser, now that the Army had got bogged down in France and Belgium, then in January took *U21* out again on a longer and more ambitious voyage. Drake had once singed the King of Spain's beard, now Hersing was going to have a tug at King George's.

First he had to thread the minefields in the Channel again, with the fast destroyers of the Dover Patrol criss-crossing the area looking for periscopes.

There were one or two near-misses, but he made it through the Straits of Dover on 22nd January, into the Western Approaches and round Land's End. The Admiralty had not yet made any of the improvements in the defences of the Straits called for by the torpedoing of the *Formidable*.

There were some tempting targets amongst the mass of merchant shipping in the St George's Channel but he held back, with his special objective in mind, and crept past the Welsh coast into the Irish Sea.

The first England knew of him was when *U21* surfaced in Morecambe Bay off Walney Island. Gunners of the nearest coastal battery took her for a British submarine, but the Navy reported none in the area. Naval experts from Vickers' shipyard in Barrow were examining her through their binoculars when they saw a flash from her gun, and shells began to fall on the airship sheds.

Equally astonished was the skipper of the 3,000-ton SS *Ben Cruachan* when *U21* rose from the sea fifteen miles north-west of Morecambe Bay and fired a shot across her bows. The ship stopped, and Hersing read her papers. She was heading north for Scapa Flow with coal for the Grand Fleet, once more at anchor there. As her crew rowed for the Lancashire shore, Hersing had explosive charges placed in her bunkers, then hauled off and watched as her sides blew out and Jellicoe's coal sank to the bottom of the Irish Sea. The *Ben Cruachan* was followed by the *Linda Blanche* and the *Kilcowan* and topped off with the sinking of the *Estacha* eighteen miles north-west of the Bar Lightship off Liverpool. Destroyers and patrol craft then made the scene too hot, and Hersing returned to Wilhelmshaven, having given north-western England its biggest scare since John Paul Jones had burned a ship in Whitehaven harbour in 1778 and spiked the guns of the fort.

In his destruction of merchantmen Hersing had always followed international law, and showed every consideration for the crews. On one occasion he even commandeered a fishing trawler to help rescue some of the merchant seamen.

Men like Hersing might continue to act under the law, but other captains, taking German submarine policy into their own hands, would sooner or later pre-empt unrestricted warfare, bearing out Bethmann-Hollweg's original fears.

The tall, broad-shouldered, blond-haired, blue-eyed bachelor Kapitänleutnant Walter Schwieger of *U20* came of an old Berlin family. Well-educated, poised, urbane, witty and courteous, he ran a disciplined but happy ship and was popular with its officers and crew and colleagues in the U-boat service. He did not support the idea of sinking without warning, but he was a careful, cautious commander, unwilling to expose his boat and crew to unnecessary danger. If there was any doubt about the warlike nature or intention of a potential target, Schwieger would be inclined to fire first and check the facts later. His copies of *Jane's Fighting Ships 1914* and *The Naval Annual 1914* listed a large number of British ships described as auxiliary cruisers or armed merchantmen, and he was not to know that most of those, though intended for conversion from merchantmen and fitted with gun rings under their deck planking, had not been armed after all.

Schwieger had spent most of December on a boring and fruitless cruise up the Danish coast, and on 26th January had left the Ems River for the English Channel.

The 3,912-ton freighter *Tokomaru*, built at Wallsend in 1893 and owned by the Shaw Savill & Albion Steamship Company of Leadenhall Street, London, had left Wellington, New Zealand, on 9th December for Le Havre with a cargo of frozen lamb. She touched at Montevideo on 2nd January and spent a day at Teneriffe in the Canary Islands on 22nd January loading fruit and more general cargo. There were many Germans at Teneriffe, and boats crowded round the ship. It was easy to find out what sort of vessel she was and where she was heading.

A French warship spoke her off Ushant and asked for identification and destination, and at 9 a.m. on 30th January *Tokomaru* was about seven miles from the Havre lightship, with the master, Francis Greene, and the second and third mates on the bridge, when there came a jarring shock and an explosion on the port side. The sea shot up over the bridge, drenching everyone there and filling the chart room. Instinctively they all looked to port and the sea started to break over her. Captain Greene got his crew of fifty-seven into the boats. He tried to save the ship's papers but his cabin was full of water. The boats hit the water, and as soon as they saw the foc'sle head go under they rowed away from the ship, and were taken aboard the French minesweeper *Saint Pierre* half an hour later. They stood by on deck until *Tokomaru* sank at 10.30. The crew were all safe but had lost their gear, and 6,000 tons of freight, mostly foodstuffs, had gone to the bottom. Because of her main cargo Schwieger was afterwards known in the U-boat mess at Kiel as 'the mutton killer'.

Two hours after the *Tokomaru* had gone down, the 2,828-ton SS *Ikaria* of the Leyland Shipping Company, Liverpool, had just stopped about twenty-five miles north-west of Havre, to take a pilot aboard, after a voyage from Rio de Janeiro, Bahia, Pernambuco and Madeira with 5,000 tons of coffee, hides and general cargo; two tugs had almost come up with her, flags at the fore, when Captain Matthew Robertson and the chief and second officers, who were on the bridge, saw the wake of a torpedo about thirty feet away, heading for the ship on the port side. The torpedo hit her for'ard. There was a loud explosion, and a mixture of sea water, coffee, cement and parts of the torpedo were hurled about sixty feet in the air and fell over the deck. The ship immediately began to sink by the head, but an hour later she was still afloat. Ropes were passed and the tugs towed her into Le Havre.

Both ships had been unarmed, both had been attacked by *U20* without any warning. And by this time Schwieger had sunk another vessel. The British steamer SS *Oriole*, a new ship belonging to the General Steam Navigation Company, had left London on the 29th. Due at Le Havre next day, she had not arrived. She had spent Monday to Friday at Norwood's Buoy loading her 1,535 tons of cargo, which had included meat, rice, tea, rubber, lead and a large amount of clothing, all for British troops abroad, and had been taken down river by Mr Fishenden, Trinity House pilot, who generally did the company's work, and whose son, F.C. Fishenden, was Chief Officer of the *Oriole*. The pilot had left the ship at 3.30 p.m. on Friday the 29th just below the Ovens Buoy, and Captain Dale's final remark to him had been, 'I expect to be in Havre early tomorrow.' She never arrived. On 6th February two lifebuoys marked SS *Oriole* were picked up on the Kent coast, and on 22nd March a stained envelope embossed on the flap 'General Steam Navigation Company, London' was found near Guernsey. The words '*Oriole* torpedoed' were scrawled on the front of the envelope. These pathetic clues confirmed that a crew of twenty-one men, a ship worth £40,000 and a cargo worth £70,000 had been lost.

All three ships were the victims of *U20*, and on the Sunday night, 31st January, Schwieger tried again.

It was a clear night twenty-five miles north-east of Cape Antifer on the French coast, though there was a strong south-west wind and a heavy sea running, and Schwieger saw the ship heading south-west, silhouetted against the moonlight which broke at intervals through the cloudy sky. She was only a tiddler, and he did not want to waste a torpedo on her, so decided to come up and deal with her.

Up above, the little 468-ton coaster *Lisette*, bound for Honfleur from Goole, Dorset, was butting head-on into the wind and sea. She was shipping too much water for'ard for 21-year-old Ordinary Seaman Arthur Brumby, at sea since he was fourteen, to keep lookout on the foc'sle, and ever since he had come on watch at eight o'clock he had been up on the navigation bridge. The ship was at full speed, the second officer was on watch, Able Seaman Waterhouse was on the wheel in the open on the upper bridge. At ten o'clock Brumby and Waterhouse changed round, and half an hour later the second sent Waterhouse to report a light to the master. He had just left the bridge when Brumby felt 'a heavy shock for'ard as though we had struck something. The vessel shuddered from stem to stern and she heeled over to port a good bit.'

The *Lisette* rolled back over to starboard, then her head dipped into the heavy seas. The second mate immediately stopped engines and ordered Brumby to call all the hands out. Captain Tom Waterhouse came up on to the bridge. Brumby was going down the ladder on the starboard side when a dark shape almost under their starboard bow caught his eye. Brumby had spent some time in the Navy and he knew what he was looking at.

'Look! There it is! It's a submarine!' he shouted up the ladder.

The submarine seemed to have been heading across their bows from starboard to port. 'Her stem,' Brumby reported, 'was lifted right up out of the water as though the fore part of the submarine was under our bow. She was in sight less than a minute when she sank and my vessel righted. I saw part of the after deck of the submarine which is raised and I also saw one fin and the propeller race.' He saw no lights or men on her deck. She seemed to be painted a whitish colour. He 'took her to be a German submarine trying to get across our stem or dive under us to leeward to speak us and that she had either misjudged the distance or not dived deep enough and had come across our stem.'

He was right. In the darkness Schwieger had misjudged the coaster's speed. With a badly damaged periscope and water coming in at the joints every time *U20* dived, Schwieger made off, to report that he had been the victim of belligerent ramming tactics by a merchant vessel. The boat was not damaged too much to prevent his firing a torpedo, again without warning, at the hospital ship *Asturias* on 1st February, even though she was painted white and carried brilliantly illuminated red crosses on her sides. According to Schwieger he had acted entirely logically. As the *Asturias* was outward bound from England, she could not have been carrying wounded, and was therefore fair game.

At this time von Pohl had the text of an announcement of all-out war on Allied merchantmen, approved by the Chancellor but not agreed by the Kaiser, all ready, and was determined to see it ratified. To bring this off he knew he would have to bypass Admiral von Müller, a cultured, artistic, unwarlike officer who considered the scheme barbaric, and who, as Chief of the Naval Cabinet, was the Chief of Staffs' link with the Kaiser.

On 2nd February von Pohl and von Müller both attended the Kaiser on an inspection of the fleet at Wilhelmshaven. Von Pohl made his request directly to the Kaiser, who gave his agreement.

On 22nd February Stoss's *U8*, an old kerosene boat, ushered in the new campaign by sinking five ships without warning off Beachy Head. Weddigen, destroyer of the Live Bait Squadron, in his new *U29* sank four ships off the Scilly Isles on 12th March, though giving their crews plenty of time to escape in the boats, as he always did with merchant ship victims. Then *U34* sank four ships without warning off Beachy Head. When he was east of the Pentland Firth on his way home, Weddigen attacked the 1st Battle Squadron, and *U29* was rammed and sunk by HMS *Dreadnought*.

Twenty-eight Allied merchantmen were sunk by U-boats in March. Still there was no effective method of sinking a submerged submarine, though the British were tinkering half-heartedly with a 'cruiser mine' which could be dropped from a ship or an aircraft. The drifter *Roburn* did catch *U8* in her anti-submarine nets, and the U-boat was sunk by gunfire from the destroyer *Ghurka*. When they picked up Stoss and his crew, the Navy did not realise that they had one of the worst breakers of the Cruiser Rules in their hands, though the U-boat sailors were at first thrown in jail as criminals anyway, rather than prisoners of war.

On May Day there steamed out of New York harbour the great Cunard liner *Lusitania*, bound for Liverpool. In addition to a crew of 700 she carried over 1,200 passengers, 440 of them women and children, 159 American, including the millionaire Alfred Vanderbilt. On 22nd April some American papers had carried a notice from the German Embassy warning Americans against sailing in Allied ships. Some people still left their luggage unpacked in their cabins and Captain Turner had been expecting his voyage to be delayed or cancelled.

There might well have been a mass disembarkation if these nervous travellers had known what the ship was carrying in her holds forward of the boiler rooms. Her cargo was almost all contraband of war, including such items as copper ingots and brass rods. Most suspect and dangerous of all were 1,248 cases, weighing fifty-one tons, of shrapnel shells, 4,927 boxes of ·303 ammunition, further undeclared cargo which was probably more ammunition, and a mysterious consignment marked on the manifest as '325 bales of raw furs', thought later to have been gun cotton of the pyroxyline type, liable to explode on contact with sea water.*

The building of the *Lusitania*, and her sister ship *Mauretania* had been financed by a loan from the Admiralty, which also subsidised running costs, conditional on both ships being available for conversion into auxiliary cruisers on outbreak of war and on giving priority to Admiralty over commercial cargoes at all times.

On her maiden voyage in 1907 she had won the Blue Riband with a speed of nearly twenty-six knots. Though for economy's sake Cunard had had Number 4 boiler room closed down, she would still be relying on her speed to rush her through U-boat Alley off southern Ireland, where she was to rendezvous with the old cruiser *Juno*, based on Queenstown. *Lusitania* had been withdrawn from service in the summer of 1915 to have her decks strengthened, magazines fitted, and rings let into the decks to take twelve 6-inch guns, but she had never actually been armed. She was still officially a merchantman – though the Admiralty used their shippers' priority to transport war stores in her. However, in *Jane's Fighting Ships 1914* and *The Naval Annual 1914*, which all U-boats carried for identification purposes,

*v. *Lusitania* by Colin Simpson.

she was listed as an auxiliary cruiser.

Von Rosenberg-Gruszczynski's *U30* sank a collier off the Wolf Rock on 3rd May, and was sighted near Fastnet Rock. *Lusitania* was ordered to clear Fastnet 'by at least ten miles'.

At the same time Walter Schwieger in *U20* was off the north-west coast of Ireland on his way south to round the Fastnet Light and head up into the Irish Sea. She made good speed on the surface, and at 9 a.m. the following morning, 5th May, she was sighted just to the north-west of Fastnet.

If this U-boat, the Admiralty calculated, remained on station in that position, she would meet the *Lusitania* at dawn next morning, with *Juno* somewhere near but not necessarily in position to give protection to the liner. Belatedly, the Admiralty decided that the old cruiser, unescorted, was too vulnerable to submarines, and ordered her back to Queenstown, meaning to replace her with destroyers from Milford Haven.

Captain Turner in *Lusitania* was warned of the presence of U-boats off Fastnet. With this and the threat of fog in mind he decided to round Ireland twenty miles further to the south and head for the middle of the narrow St George's Channel between Ireland and Wales.

In *U20*, just south-east of Fastnet, Schwieger, also heading for Liverpool, had likewise set a course twenty miles off the Irish coast, identical to Captain Turner's. If he continued on it the *Lusitania* would overhaul him. Then, at 5.30 p.m. he sighted the sailing schooner *Earl of Lathom* and sank her with explosive charges. At 7 a.m. on the morning of the following day, 6th May, when *U20* was approaching the mouth of St George's Channel, she sighted and sank the SS *Candidate* in the gap. Because *Candidate* had been armed, when Schwieger intercepted her sister ship *Centurion* early in the afternoon he torpedoed her without warning.

Schwieger then had to reassess his position. Persistent fog would seriously hamper his operations in the Irish Sea and make it much more difficult to avoid patrols there and in the St George's Channel, and he had used up more fuel than anticipated on the passage south. He decided to abandon his mission to the Mersey and turn west for the Atlantic.

That evening Captain Turner received a wireless message from the Admiralty warning him of U-boats off the south coast of Ireland. He waited for another signal instructing him to turn northabout up the western coast of Ireland or make for Queenstown, but none came. Then he thought that *Juno* would bring him the order when he came up with her at dawn. He did not know that she had been ordered back to Queenstown.

Next morning, Friday 7th May, the fog was thick. Turner slowed down to fifteen knots and started sounding his foghorn.

At 11.52 he received an Admiralty signal warning him of a U-boat in the middle of the St George's Channel. This signal was badly out of date. The U-boat had retreated to the west since then and she and *Lusitania* were now closing each other, some hundred miles apart. Turner increased speed to eighteen knots, as the fog cleared.

About noon Schwieger heard powerful screws overhead. He raised the asparagus and saw the *Juno* as she made for Queenstown. A few minutes later he sighted a big four-funnelled steamer, and identified her from his reference books as either the *Lusitania* or *Mauretania*. The big liner was four hundred yards away on his starboard side when he gave the order to fire.

Londoner Oliver Bernard, who had once sailed before the mast, was standing on the starboard side of *Lusitania*'s Palm Lounge looking out to sea when he saw 'a streak of white on the water that looked like the tail of a fish'. He 'watched it for a minute and made out the periscope of a submarine and then saw the wake of a torpedo making for the *Lusitania*.' At that moment a woman passenger came up to him and said with a laugh, 'Looks like a torpedo coming.'

Edward Berghway, able seaman and saloon deck man, saw the periscope feather three hundred yards abeam, then the wake of the tinfish. He sang out a warning, then made for his boat, on the starboard side.

There were double lookouts in the Crow's Nest. Tom Quinn had taken over port side lookout at two o'clock with a cheerful 'Anything in sight?' 'Nothing doing,' said AB Seagrave, handing over. Ten minutes later he saw the white wake two hundred yards off, making about thirty-five knots and ranging for their foremast.

He said to port look-out Hennessy, 'Good God, Frank, here's a torpedo!' then shouted down very loudly to the bridge.

'Torpedo coming to strike us amidships!' Second Officer Heppert sang out, 'Here is a torpedo!' Then it hit them.

American C.T. Hill, Director of the British-American Tobacco Company, talking to Chief Steward Jones in the entrance to the dining saloon, heard the sound of impact as 'like wind slamming a heavy door'. Falling debris made him cover his face with his hands. Leading Fireman David Evans, on the port side of Number 3 stokehold, heard 'a big crash as if the vessel had struck a rock'. The stokehold filled with dust, and a man shouted, 'They've got us at last!'

Schwieger had over-estimated his target's speed, and his torpedo hit the *Lusitania* just forward of the bridge at one of the most sensitive points on the entire hull.

Their financial stake in the *Lusitania* had allowed the Admiralty to impose their own engineering layout. So-called 'watertight' compartments built in outboard of the boiler rooms were in fact used as bunkers, with hatches open to the boiler rooms. The ship was also top heavy. Cunard had built the biggest, most luxurious floating hotel in the world, five decks crowned by heavy superstructure and four towering funnels. The *Lusitania* was like a huge racing shell with an overweight oarsman and a bomb in her bows.

The torpedo struck the spot where the athwartship bulkhead separating the starboard 'buoyancy' bunker from the lower orlop hold joined the ship's side. Schwieger, like most of his fellow U-boat captains, had had constant trouble with dud torpedoes, but this one exploded, and the blast blew not only back through the bunker and open hatches into the boiler rooms, but forward into the hold. From the water filling the bunker, the towering, top-heavy ship took up a rapid fifteen-degree list to starboard. The explosive gases flashing into the hold ignited some part of or perhaps all the contraband explosives stored there. There was a second, deep, rumbling explosion which blew out the whole bow below the waterline.

The cloud of smoke, white damp and coal dust which shot up through the bunker ventilators on deck at the time of the second explosion, and the line of bubbles which was actually the compressed air reaching the surface from the first torpedo's motor, confused some passengers and crew into imagining a second torpedo.

The great ship began an inexorable nose-dive, her stern rising up out of the water. Turner ordered 'Abandon ship' and told the Marconi operator to send SOS signals.

The ship's foredeck was now completely under water, and the sea pouring in through the forward hatches. Further aft, water gushing in through open portholes had reached D deck. Turner gave a wheel order to try to turn towards the Old Head of Kinsale, but the rudder was now clear of the water and there was no response.

Through his periscope Schwieger was watching in shocked fascination the panic and muddle on the boat deck. With the ship heeled over so far to starboard, only the boats on that side stood any

Light cruiser HMS *Hawke*, sunk by *U21*, 15th October 1914.

Attack on the grounded *E15* in the Dardanelles by Royal Navy picket boats, on the night of 18th/19th April 1915.

(*Left*) Lieutenant Martin Eric Nasmith, VC, CO of HMS M *E11*. (*Right*) Lieutenant-Commander Norman Holbrook, VC, Commander of Submarine *B11* which torpedoed the *Messudiyeh* in the Dardanelles.

Nasmith's *Ell* being cheered as she returned from her first successful patrol in the Sea of Marmara 1915.

chance of getting into the water, and even these were swinging out so far that it was difficult for the passengers to get into them.

The ship was tilted so far over to starboard that no boat on the port side could be lowered directly into the water. Any boat which was lowered past boat deck level grated down the ship's side, tearing its planking to pieces on the rivet heads. One boat swung inboard, crushing the people who had gathered in the collapsible boat beneath it. The after fall of another boat parted, spilling its passengers into the sea, then fell on to those in the water. Many people simply jumped into the sea.

At 2.30 p.m., the *Lusitania* disappeared, just twenty minutes after the torpedo had hit her. The suction pulled down and drowned many of the people swimming or struggling in the water, as well as others still aboard, injured, dazed or trapped below.

When Admiral Coke at Queenstown received *Lusitania*'s SOS just after 2.15, he ordered the *Juno*, which was in harbour with steam up, to sail at once to her help, and she was now on her way. She had just cleared harbour when another message came in, from the lighthouse keeper on the Old Head of Kinsale, reporting the liner sunk. Coke immediately ordered out all available craft.

Only six lifeboats and a few collapsibles floated out of *Lusitania*'s forty-eight boats. Leaving his own boat trailing uselessly from its after fall, AB Berghway had pushed further aft along the starboard side through the milling crowd of passengers to Number 15 boat, and helped First Officer Arthur Jones to fill it, mainly with women and children. They packed about eighty people in the boat 'with great difficulty owing to the distance between the ship's side and the gunwhale and lowered it in safety into the water,' Jones reported. 'The falls were cast off and the boat commenced to drift astern close to the ship's side. The forward part of the ship was submerged and the forward bridge awash and people were slipping down the deck into the water.' People in the boat shouted up to Jones asking him if he was coming down to join them. He looked round and seeing it was impossible to do anything more slid down the falls and was pulled into the boat. The boat continued to slip astern and about three minutes later, as Berghway related, 'The *Lusitania* took a tremendous dive, practically standing on her head, and sank killing hundreds of people.'

AB Frank Hennessy was washed off the boat deck. In the water he clung on to a boat's chock and gave it up to a lady passenger; he managed to swim to a collapsible boat with several people in it. The boat repeatedly turned over, losing some of its passengers each time, then the bosun's lifeboat appeared and took the survivors aboard.

Third Officer John Lewis had grabbed the lifeline of a collapsible boat as the water overtook him and clung to it. He came to the surface to find that the ship had gone. He joined twenty-five people clinging to an overturned boat. When the bosun's boat came up with them there were only five left hanging on.

Fireman David Evans was in Number 17 starboard boat when the fall slipped and nearly everyone was thrown into the sea. He hung on to a boat seat but the ship took the boat down with her. He struggled to the surface and caught hold of a rope attached to a collapsible boat, floating right way up with its canvas cover still on and two people sitting on the top. He recognised Trimmer McKenna, who was helping a lady passenger to hold on. McKenna helped him up, then they cut the cover off, put the sides up and, baling all the time, picked up about fifty people from the water.

Admiral Coke had signalled the Admiralty reporting the sinking and his rescue operations. When Lord Fisher read the message at about 3 p.m. he immediately ordered the *Juno* recalled for the second time, fearing another 'live bait' sinking. When the cruiser received the order to turn back she was actually within sight of the *Lusitania* survivors, but she obeyed the command and reversed course. It was another two hours before the first small craft reached the scene, and more people drowned in the interval.

The fishing smacks and naval patrol boats searched all night for survivors. When the morgues of Queenstown and Kinsale had received their last corpses, out of 1,978 men, women and children who had sailed in the *Lusitania*, a tally of 1,198 dead was counted. Ninety-seven of the dead were American citizens, and one of them a Vanderbilt.

The British newspaperman Hayden Talbot took a cab to work in Fleet Street on the morning after the sinking, and at Temple Bar he gave a lift to an American colleague. They talked about the subject that was on everyone's lips.

'Thank God it's happened!' said the American excitedly.

'Thank *God*?'

'Yes, thank God! Now we've *got* to be men and fight!'

Vance Pitney, London Staff Correspondent of the *New York Tribune*, cabled 'Germany has committed her latest crime of savage inhumanity'; other Americans in London and the British public expected the United States to declare war on Germany.

But the time was not yet ripe. Americans back home were not unanimous in demanding war. The Eastern press, churning out lurid stories by survivors that the U-boat captain had come up to fire a whole salvo of torpedoes into the sinking liner, even that he had fired poison gas shells, thundered their righteous, jingoistic wrath, but America was slow to anger, especially in the mid-western states, where there were many German-American settlers, and where isolationism was deeply rooted. Wilson also needed the German-American vote in the next Presidential election in 1916.

Walter Schwieger brought *U20* back to Germany, where his fianceé found him, 'So haggard, so silent, and so *different*,' and he was reluctant to appear in public. With the obliquy of the Allies and of the USA upon his blond head he went back on operations, and was drowned in *U88* on 17th September 1917.

CHAPTER FIVE

E-boats and the Iron Trade

Max Horton's *E9* had found a dearth of targets, though she did sink the destroyer *S116* on 6th October. After her sinking of the cruiser *Hela* on 13th September, the German High Seas Fleet retreated down the Kiel Canal and exercised only in the Baltic, where there was already a mass of German shipping, including troopships heading for the Russian front and their escorts, which had little to fear from the Tsar's ramshackle, badly disciplined fleet. A few well-handled British submarines, on the other hand, would find plenty of targets in this 'German Lake'.

In the dark early morning of 15th October three E-boats, Horton's *E9*, Martin Nasmith's *E11*, with Lieutenant-Commander Noel Laurence in *E1* commanding the squadron, lay in Gorleston harbour, near Yarmouth on the Norfolk coast, preparing to leave for the Baltic. At the last minute Nasmith signalled that he was having engine trouble and would not be ready on time. Laurence ordered him to follow as soon as possible, and at 5 a.m. led the remaining two boats out of harbour and north-east across the North Sea. They had only reached Smith's Knoll when Laurence signalled Horton that he too had had a mechanical breakdown and intended to stop for repairs while *E9* carried on independently.

The lone submarine proceeded on her gas engine, spray breaking over her bridge, rolling and pitching with the swell. At 10.30 a.m. next day, 16th October, they sighted Boobierg Light, Jutland, and steered a northerly course along the Jutland coast. Three times they had to dive and stay down for a while to avoid being seen by shipping. At 6.25 p.m. they sighted what looked very like the wash and wake of another submarine, and remained dived for half an hour. At midnight *E9* rounded the Skaw and headed south-east across the Kattegat towards the Swedish coast.

At dawn, about 4.30 a.m., she dived, with merchant shipping no more than two miles off, and ran submerged until 7.30 a.m. Horton came up cautiously. At eight a freighter got within sighting distance and at 9.15 a.m. the sub dived again when she sighted a Swedish cruiser, and stayed down until dark, with the cruiser or other thick traffic always in easy sighting distance. She surfaced at 5.45 p.m. and made for the Sound, the passage into the Baltic between the Copenhagen shore of Denmark and the Swedish coast, threading a continuous stream of traffic coming in both directions. At 11.28 p.m. *E9* was three miles north of the Kullen Light near the entrance to the Sound. There was not enough time to clear the Sound before daybreak, with all the patrols, neutral and German, infesting it; Horton took it it for granted that both the other submarines had been similarly delayed by the need to stay unseen, so he dived to the bottom to wait for nightfall on the following day.

He was wrong about the progress of both the other boats. Nasmith's *E11* was still in Gorleston harbour, but Laurence had overtaken him. Her repairs made, *E1* left Smith's Knoll at 10 a.m. on the 15th and making a good ten knots across the North Sea reached the Skagerrak late on 16th October, rounded the Skaw and steered into the Kattegat, submerging whenever shipping got too close but making better time through the traffic than *E9*. Laurence arrived off the northern entrance to the Sound at 8.30 p.m., some three hours ahead of Horton, submerged and began the passage through as soon as it was dark, and by 11.30 p.m. was safely through. A German patrol was seen burning searchlights ahead, and as he did not want to find himself at dawn with exhausted batteries he did not dive past them but went to the bottom, at just about the time Horton was doing the same, only at the wrong end of the Sound.

At first light on the 18th Laurence rose to surface, took a cautious look first through the periscope, and had a close-up view of a German destroyer. He stayed down and steered to the southward, towards the German shore, and at 9.10 a.m. sighted smoke. The smoke became masts, the masts the two-funnelled protected cruiser *Victoria Luise*, an old ship of 5,885 tons launched in 1898 and used mainly as a training ship, but a worthwhile target, with her two 8.2s, six 5.9s and top speed of nineteen knots.

She was heading in their direction. Laurence shaped course to intercept, and at 10.6 a.m. she came abeam at five hundred yards. He fired two torpedoes, with an interval of one minute. The *Luise* had keen lookouts, who spotted the white tracks heading for them. The cruiser started turning away, though everyone on deck saw grimly that it would be too late to avoid the first tinfish.

Luckily for them, British submarine captains were still not allowing for the weight of a warhead, and the torpedo passed under the ship and did not explode. She had her helm hard over, the second just missed her bows, and Laurence had to dive to seventy feet to avoid being rammed.

The panic was on. In Kiel destroyers, torpedo-boats and armed trawlers were ordered to the Sound, off the entrance to which lay *E9*, waiting to make her dash through.

E1 surfaced at 10.30 a.m., saw no sign of her previous target but some ten minutes later sighted a small cruiser, decided to attack, but could not get within range. At 11.30 she sighted another protected cruiser of the *Hertha* class, probably the *Victoria Luise* again, making frequent big alterations of course every few minutes, as if carrying out a search. Laurence stalked her for six hours but could not get near enough to attack. She opened up with one brief burst of fire from her main armament, though not in their direction.

E9 lay on the bottom off the Kullen Light all that day, the 18th. Hands were piped to dinner in the normal way, Horton had his habitual rubber of bridge with his officers in the tiny wardroom, punctuated by two cautious ascents to periscope depth for a look round, though Horton dared not surface to ventilate the boat. He saw many merchantmen but no warships. By 5.20 p.m. it was quite dark enough to surface and creep forward on the gas engine through the Sound, with the fans pushing sweet fresh night air through the boat, cutting the stale, sour effluvia of men and machinery.

Scraping shoals and steering carefully through the narrows between Danish Helsingor and Swedish Hälsingborg, *E9* passed Malmo Light on the Swedish side at 11 p.m. Trimmed down, with the upper deck just awash, the sub went through the Klint Channel slowly, on one engine. 'Night calm, clear and dark,' wrote Horton appreciatively in his log.

He kept as far to the north as possible to avoid destroyers. At eleven o'clock the Drogden Light Vessel was abeam. Half an hour later, with the lightship some three miles to the north-east, a destroyer suddenly loomed 150 yards on their starboard bow . . .

'Dive dive dive!'

E9 got under in fifteen seconds, timed by stopwatch. At fourteen and a half feet by the flickering needle on the depth gauge there was a shuddering crash as she hit bottom. Horton stopped engines, peered at his chart and pencilled a cross on the four and a half fathom patch where he reckoned they must be. The bottom here was very stony, and the tops of their twin periscope standards were sticking out about six inches above the water.

After about quarter of an hour he came up and saw the destroyer seventy yards on their port beam, dived again and went ahead at ten feet, leaving just four feet at best between *E9*'s keel and the bottom, and steered to pass the enemy vessel, which had apparently not sighted them. Any rocky outcrop on the bottom could rip the sub's plates as if she were a rowboat. But they left the destroyer thankfully astern and slipped into deeper water, fifteen feet, then twenty, and at 3.40 a.m. on the 19th, with Steons Klint three points on their starboard bow, they rested on the bottom at forty-eight feet. At five o'clock they rose and went ahead submerged with the periscope revealing destroyers ahead and astern. At 10.16 a.m. they surfaced but had to dive again for a destroyer coming up fast from the south-west, the direction of Kiel. They heard the rumble of her engines as she passed close. At three o'clock in the afternoon the sea through the periscope looked bare so they surfaced and chugged east towards the open Baltic, diving at 5.30 after dark to lie on the bottom on the Moen Bank off the Danish shore for the night, with a good charge in the batteries ready for the patrol into the Gulf of Danzig which had been scheduled for the squadron's first formal operation.

Laurence in *E1* had spent the whole of that day off the southern coast of the island of Bornholm, eighty miles east of Moens into the Baltic, saw no likely targets, and next day, 20th October, headed east for the Gulf of Danzig. By afternoon *E1* was inside the harbour. Laurence sighted three cruisers in the inner basin but could not get near them.

E9 rose at 5.30 a.m. on a frosty morning and steered east, passing Cape Arkona on the German coast eight miles to starboard. There was no sign of patrols. At 1.25 in the afternoon they were skirting the southern edge of the Ronne Banks off Bornholm when they sighted a cruiser to the north. Wind and sea increased steadily. At 10 p.m. they dived. At one o'clock on the morning of the 21st insulation on the starboard main motor started to burn. *E9* came up, and an inspection revealed a short in the winding. Horton gave up the idea of Danzig and steered on the gas engine for their new base at Libau (later Liepaja) on the Russian Lithuanian coast. They ploughed through heavy seas and early on the 22nd sighted tugs off Libau. The tugs took them over and Russian naval officers piloted them into harbour, where they found *E1*, but no *E11*. Both *E1* and *E9* had passed unaware through several minefields off the port, which it was feared would shortly fall to the German. The dockyard and plant had already been destroyed. The two British submarines stored, made repairs and rested their crews while they waited for *E11*.

They would wait in vain. Nasmith had reached the entrance to the Sound and tried to get through on the night of the 19th. Twice enemy patrol craft had sighted him and he had crash-dived, knowing very well that the German Navy would not hesitate to ram him in these neutral waters. A verdict of 'accidental collision' would be a perfect excuse in narrow, crowded roadsteads like these. He hoped to try again on the night of the 20th but was forced to come up during daylight to recharge his batteries and ventilate. He was careful to surface in an area temporarily free of surface hunters, but had the bad luck to encounter the other new invention which was in later years to be paired with the U-boat as a 'weapon of the weaker nation', then to be turned against it as the most effective anti-submarine countermeasure. In one of the very first examples of co-operation between aircraft and surface warships against submarines, a seaplane spotted *E11* on the surface and called up destroyers. Nasmith was not a man to be easily defeated, but he was forced to turn back and return to Harwich.

On 24th October there was a message from the British Ambassador directing *E1* and *E9* to go north immediately to a new base at Lapvik on the southern coast of Finland, but Laurence wanted to give the men more time to rest and allow Nasmith a little longer to arrive, and postponed departure for another twenty-four hours. When he did leave on the 25th, in company with *E9* and the Russian submarine *Crocodile*, it was to have one more try for the Danzig cruisers first, as he knew that once in Lapvik both his boats would be out of action for some weeks refitting. Twice *E1* went into the harbour, fired at and missed a new destroyer on trials, then Laurence sailed for Lapvik via the west coast of Gotland where he looked unsuccessfully for German cruisers on their regular north-south Baltic route. On 30th October he rendezvoused with *E9* and the two boats were guided into freezing Lapvik (Turku) by a Russian patrol boat. Next day Laurence was taken to Helsingfors (Helsinki) on the Gulf of Finland by a Russian destroyer to report to his new C-in-C, the Russian Admiral Kanin and his Commodore Submarines, Podgoursky, and was directed to take his two boats across to Reval (Tallinn) in Esthonia for their refit.

With new engine clutches they were ready for action by 13th November. On the 17th south of Bornholm *E9* attacked a 2,000-ton *Gazelle* class cruiser from five hundred yards but both torpedoes fired broke surface directly after firing; the enemy sighted them and swung towards the submarine too quickly for Horton to aim with his beam tubes. The weather grew colder, the sea rougher, and a sub would find it very lively even down at sixty feet. It became increasingly difficult to get into and out of harbour for the thick ice; spray froze to solid ice all over upperworks, immobilizing periscopes, telegraphs and tube caps. The Russian subs went into winter hibernation, but the two Royal Navy boats continued patrols through November and December, looking for an elusive German squadron of two armoured cruisers, a light cruiser and a gunboat, and deliberately showing themselves to coastal lookouts and enemy ships. *E1* attacked a destroyer west of Gotland on 12th December and was nearly caught in an explosive sweep towed by a trawler. All this rough sea time meant more repairs, usually to clutches. *E9* was ready once again on 15th January, *E1* on the 23rd, to carry out Admiral Kanin's orders for offensive action against the German Fleet, supporting the flanks of its army, which was advancing rapidly into Russia.

On the morning of 24th January *E9* left Reval behind the ice-breaker *Peter the Great*, which had to remain with them to clear a passage until seven o'clock that evening, with the submarine making for Gotland and an enemy cruiser reported aground. The cruiser had gone, either sunk or towed off. Horton patrolled round Bornholm, then went south-west. At 11.30 a.m. on the 29th, between Moens and Arkona, the weather was clear, visibility good. 'As circumstances were favourable,' recorded Horton, 'decided to attempt to bag a destroyer.'

He had the torpedo depths altered to five feet, and at two o'clock in the afternoon sighted three destroyers rounding Moens Klint heading for the Sound. He closed but they were too far off. An hour later another destroyer was seen coming up from the south on the same course.

In another ten minutes she was six hundred yards from *E9*, seven points on her bow. A heavy swell made depth-keeping very difficult, but Horton fired a bow torpedo. He saw it running well, then dipped to avoid detection. The count-down reached fifty seconds. There was the unmistakable noise of a torpedo detonating. Horton rose to seventeen feet and took a look through the periscope. The destroyer had completely disappeared. A second destroyer attacked five minutes later saw and avoided him, and when a squadron of four approached zigzagging at high speed the light was too bad for a shot.

At 7 p.m. *E9* came up and headed for Bornholm in a very cold north wind and a rising sea. Spray froze as it struck and the bridge became a mass of ice. The bridge screen was immovable, with six inches of ice on it. Valves and vents were freezing up, torpedo tube caps and periscopes froze solid, and a man had to be stationed at the conning tower hatch to keep that free and prevent those on watch on the bridge from being marooned and frozen into figure-heads. Coated overall in ice, *E9* looked like a long craggy ice floe. Horton got the watch chipping away with hammers and chisels, then at 10.25 ordered 'Diving stations.' With all her topweight of ice, *E9* went down like a plunging whale, but on the bottom on Foranderly Bank the temperature was warmer, and the ice melted away.

There was still much heavy floating ice about when *E9* left Reval on patrol on 28th April. Having negotiated that she had to pick her way through unmarked minefields; then thick fog brought her to a stop, with visibility fifty yards. The north wind blew the fog away next day but brought heavy seas which cut her down to five knots. Ordered to attack minelaying destroyers off Libau, Horton just missed with two torpedoes, then fog closed down again. At 3 a.m. on 10th May a wireless message reported Libau in enemy hands, and *E9* was ordered to operate on the lines Libau-Danzig or Libau-Memel (Klaipeda) against enemy troop transports.

Next day from earliest light the horizon was full of smoke, as *E9* patrolled off the southern end of Gotland. At 6.30 Horton saw destroyers' masts to the south and dived to ambush them, then an hour later a big cloud of black smoke rose from the direction of Libau, and *E9* was presented with three cruisers and their destroyer screen convoying two large transports and one small, also with their attendant destroyers.

They drew ahead fast and altered course to the west at full speed to cross the submarine's bows to starboard. The sea was mirror-smooth and as they came closer Horton was forced to slow down to reduce his periscope feather.

At 8.5 a.m. *E9* passed through the cruisers' destroyer screen and two minutes later fired two torpedoes at the cruisers from 950 yards. They altered to port and the shots missed.

The transports were coming up about three miles astern. Horton crossed their bows to attack from right ahead and got inside the destroyer screen to gain the advantage of the bright sun shining behind him.

At 8.25 he fired a port beam torpedo at two hundred yards from the leading transport. There was no result. He passed underneath her and turned to starboard. With the same deflection and relative position he emptied a stern tube from four hundred yards at the second big 6,000-tonner, and saw it hit her just for'ard of the funnel. Under fire from small guns and with detonations from explosive sweeps going off all round, he continued to swing to starboard and fired a shot from the reloaded bow tube on the stricken ship's quarter. The torpedo ran the long 1,600 yards and Horton saw the big explosion under her bow. He went deep and steered inshore to avoid the enemy destroyers, and when he came up at 9.15 his target had disappeared.

On patrol in June Horton sighted a light cruiser patrolling in a wide circle round a convoy of four destroyers escorting a collier. He missed the cruiser with a shot from his port beam tube but a salvo from the bow tubes hit and sank both the collier and the destroyer which was alongside her.

The Germans were having to divert more and more warships from the North Sea to escort supply ships in the Baltic. The two British submarines were bedevilled by fog at the end of June and in early July. Horton surfaced just before three o'clock on the afternoon of 2nd July and, with visibility about four miles, saw two big warships with destroyers.

Both torpedoes fired hit the 9,050-ton armoured cruiser *Prinz Adalbert*, built in 1903, with four 8·2-inch, ten 5·9-inch and ten 15½-pounder guns. The two feeble British 18-inch torpedoes did not sink her but she was out of action for four months. Horton received the Order of St. George, the highest Russian military award.

E1 had been active but had had bad luck with attacks. On 30th July she fired at three transports in line ahead and hit the leading ship, which blew up.

The activities of two Brtish submarines had caused the Germans to abandon altogether an attempt to capture Petrograd (Leningrad) at the head of the Gulf of Finland, by amphibious landings on the southern coast.

Horton and Laurence had been so active and had appeared in so many different places in rapid succession that the German Navy was convinced that they had a whole flotilla to deal with.

The German High Command now decided that the Allied submarine base on the Gulf of Finland must be destroyed. Vice-Admiral Schmidt's Baltic Fleet of two battleships, four cruisers and thirty-one destroyers was augmented by eight battleships, three battle-cruisers, five cruisers and thirty-two destroyers under Vice-

Admiral Hipper, temporarily withdrawn from the High Seas Fleet to help crush two small submarines.

On the morning of 19th August Laurence was patrolling in mist off the mouth of the Gulf of Riga when he sighted battle-cruisers. He only had time for a quick shot with a beam tube, but he hit SMS *Moltke*. The British tiddler was too puny to sink a battle-cruiser, but it did flood several compartments and slow her down, and it was enough to make von Hipper retreat to Danzig, aborting the whole mission.

In the spring of 1915 the Admiralty decided to bring the submarine strength in the Baltic up to flotilla strength by sending four more E-boats through the Sound and to have four of the older coastal C-class boats towed, partially dismantled, to Archangel on the Arctic coast of Russia for transport overland to Petrograd. Lieutenant-Commander Goodhart's *E8* and Lieutenant-Commander Layton's *E13* were the first boats to try it. The Germans had greatly increased patrol strength in the Sound, and Goodhart had a rough passage, but made it to the Baltic. Layton was not so fortunate. *E13* ran aground on Saltholm Flat, in Danish territorial water. Disregarding all laws of neutrality and codes of warfare, two German destroyers came up and shelled her to destruction. Halahan's *E18* and Cromie's *E19* were held back until things had cooled down. In September they both broke through. *C26, 27, 32* and *35* were moved down the canal from Archangel to Petrograd but were held up when the ship bringing their batteries separately was sunk on passage out.

Both *E8* and *E19* went on patrol in early October, with orders to attack the traffic of German merchant shipping up and down the Swedish coast, with particular reference to the flow of iron ore to Germany.

At 5.30 p.m on 3rd October Cromie's *E19,* which was fitted with a 12-pounder deck gun, rose to stop the 5,000-ton German merchantman *Svionia*. A warning shot across her bows with the submarine's deck gun did not stop her, so Cromie opened fire on her and hit her amidships with his second shot. She then stopped and her crew obeyed a signal to abandon ship. Cromie closed and fired five more shots into her, but eventually had to fire two torpedoes at her, as it was too rough to board her to place demolition charges. The first torpedo failed to run, the second passed underneath her. With Arkona Light sending up a Brock's Benefit of rockets, Cromie had to leave the *Svionia* on a lee shore with the wind rising, then head northwards for Bornholm and shelter for *E19*.

Cromie sighted many small merchant ships on the Swedish-German trade, but the weather remained too rough to rise and examine them. On the afternoon of 9th October in very ugly weather *E19* attacked a two-funnelled merchantman but was thrown off her aim by the small torpedo boat *D6*; an hour later he fired at a big destroyer returning from the Sound at high speed and was very nearly rammed; and thirty minutes after that he unsuccessfully stalked a light cruiser coming out of the Sound in shallow water.

Next day was more profitable and brought him first the ore carrier *Lulea*, though it was too rough to use the gun and she cost him no fewer than four torpedoes before she sank. He started chasing several merchantmen at 8 a.m. on the 11th off southern Gotland and at 9.40 stopped the *Walter Leonhardt* bound from Lulea to Hamburg with ore. Her crew abandoned ship in the boats, Cromie's boarding party opened her sea cocks and put a gun-cotton charge in the after hold. The charge had a damp fuse and did not go off. Back went the boarding party and put in a dry one, and the ship sank three minutes after the explosion. Cromie stopped a Swedish steamer and directed them to pick up the crew.

At noon he chased the *Germania* of Hamburg towards the Swedish coast. Cromie saw she was heading straight for shoal water, and in strict accordance with Board of Trade regulations fired his gun to try to warn her of her danger. Not having any blanks he had to use practice ammunition but took good care that the shot fell well clear of her. In her alarm she did not notice the shoals until she had struck them and gone aground. Cromie went alongside cautiously to save the crew and try to salvage the ship, and found her abandoned. He spent an hour trying to tow her off using a coil of rope from her after hold but she would not budge and with the water gaining in her engine room he had to abandon her after taking her papers and transferring her fresh meat to *E19*. Her cargo was 2,950 kilos of finest concentrated iron ore bound for Stettin from Stockholm.

At two o'clock he was after the *Gutrune*, which at three o'clock obeyed his signals to send a boat with her papers. A boarding party opened all her watertight doors and removed her main inlet value, then Cromie put three shots into her fore hold. They watched her sink reluctantly, as she was a well-found ship, elegantly fitted out, but he could not let the Germans have her 4,410,600 kilos of ore. He had just towed her boats across to a Swedish steamer when he sighted two more large steamers to the southward. In twenty minutes they were lying stopped and his men were boarding them. The Swedish *Nyland* with ore for Rotterdam was allowed to proceed but the German *Director Rippenhagen* had 950,000 kilos of ore and was sunk. The crew were put aboard the Swedish wood pulp carrier *Martha*, bound for Newcastle. At six o'clock Cromie watched his first victim of the afternoon, the *Gutrune*, sink stern-first as he closed a large steamer which steamed hard for the Swedish coast on sighting him. At 6.30 he brought her to with a shot across the bows and sent an armed boarding party, who found her to be the *Nicomedia* with 6,704,700 kilos of magnetic ore for Hamburg. A charge was exploded in her after hold and her crew pulled ashore. *E19* steered up west of Gotland, after a very busy day.

Two innocent Swedes were sent on their way early on the 12th but the *Nike*, with ore for Stettin and a Swedish master, was something of a mystery, and Cromie put Lieutenant Ince aboard to take command and bring her in to Reval in their wake for further examination. The captain of the Swedish *August*, bound for Shoreham, Hampshire, with timber, gave them fresh bread, and they turned for Reval. Captain Anderson of the *Nike* told Lieutenant Ince that twenty German ships loaded with iron ore had been held in Lulea to await escorts, as a result of *E19*'s raid. Because of her cargo of magnetic ore the ship's compass was very much out, and she grounded once on the way back, but at 7.15 a.m. on 13th October Cromie handed over his charge to His Imperial Majesty's torpedo-boat-destroyer *Dostorny* after re-embarking his prize crew. *E19* also had to alter course sharply five miles north-east of Odensholm to avoid a torpedo fired at her from long range, presumably by a U-boat.

Cromie's report of seven ore ships and some 20,000 tons of ore sunk in one patrol pleased his Admiral, and Goodhart had added the ore carrier *Margarette* on 5th October.

E9 left Reval on 17th October to replace *E19* off the Swedish coast. Next day she stopped the Danish *Karla*, registered at Esbjerg and carrying coke from Newcastle to Stockholm. Horton let her proceed, although her cargo may have been intended for the Swedish blast furnaces producing iron ore for Germany. Her master said to the boarding officer, 'First the Germans in the Sound, now the British in the Baltic. I am suddenly very popular.' He also reported seeing a German destroyer that morning.

At half-past three Horton sighted several steamers in the distance passing up and down the Swedish coast, and dived to close them. At

5.44 p.m. he let in one engine clutch ready to close one steamer on his starboard bow heading south-westerly and showing no neutral markings on her side. He ordered her to stop in International Code and fired his bridge-mounted Maxim gun across her bows, then closed her on his motors. She was the German *Soderhamn* of Hamburg bound for Delfzyl in Holland from Sweden with wood. A boarding party opened up the sea cocks and exploded an 18-pound charge in the corner of a bulkhead and the ship's side at the after end of the engine room, with her crew in the boats.

While his men were still aboard her Horton sighted a large vessel to seaward heading south without lights. The boarding party was immediately collected and with the *Soderhamn* listing and settling, *E9* chased the other ship. At 7.15 p.m. he stopped her with his signal lamp and a burst from the Maxim. Her crew abandoned ship quickly and the boats vanished into the darkness taking the ship's papers with them, but Horton managed to have a shouting dialogue with her chief officer, who established her as the 7,000-ton *Pernambuco* of the Hamburg-Amerika Line, bringing 3,500 tons of iron ore from Lulea to Stettin. Horton again put charges in her and stood by to watch her sink. After two hours she had settled very low in the water but had stopped sinking.

Horton had just sighted a suspicious craft to seaward which looked very like the German TBD reportd by *Karla*'s captain so he decided to finish the *Pernambuco* with a torpedo. Before his sights were on he ordered 'Stop' to the engine room, but the torpedo officer thought he had said 'Fire' and let one go prematurely, which missed. The second hit amidships and she went down immediately. Horton let in both engine clutches and steamed off to charge batteries, ready for some more action when daylight came.

The Baltic winter nights were long, and it was still dark at 6.40 a.m. when he sighted a steamer. He gave chase on motors and in fifteen minutes had stopped her with lamp and maxim. Her master was very forthcoming and told the boarding officer that she was the German *Johannes Russ* going north to Sundavall with coke, and then to Lulea to load ore for Germany. Horton had only small charges left by now so these were used to blow out the ship's sea cocks. Just before eight o'clock he dived to the south-eastward and saw the *Soderhamn* on the rocks with a big list, and an hour and a half later observed a destroyer standing by the *Johannes Russ*, still afloat, while he was heading submerged for another merchantman to the northward.

E9 surfaced just off her bow at 10.47 a.m. Horton closed her.

'What ship are you?'

'*Ich bin Dal Alfren.*'

'What nationality?'

'*Deutsche.*'

The officer of the watch had his binoculars trained on the *Johannes Russ* and her attendant destroyer.

'Sir, she's coming for us.'

The destroyer was heading straight for them, bow wave foaming. It was impossible to tell her nationality, end-on.

Horton fired a torpedo at the *Dal Alfren*. It malfunctioned and sank. He fired a second. It swerved to starboard for forty yards, then ran straight, just missing the merchantman's stern. The destroyer was now about a mile away. Horton took the boat down and awaited developments. As the destroyer closed he recognised her as Danish. He watched while she took the *Dal Alfren*'s crew on board, then rose and closed her.

'I am a British submarine.'

'I am *Wale*. You are in Swedish neutral waters.'

'I make myself six miles from the nearest land.'

'I make you five.'

'The neutral limit is *three* miles. Please stand clear while I sink this ship.'

With *Wale* a hundred yards on her beam, he fired a stern tube at the *Del Alfren*. This one did not embarrass him in the eyes of the Danish Navy, and the ship sank in two minutes.

On the following day, 20th October, *E9* chased and closed five ships, all Swedish timber carriers, and it was soon apparent that the operations of the two submarines had stopped all German traffic to Sweden.

The E-boats returned to the Libau patrol once again. At dusk on 23rd October Goodhart's *E8* arrived off the port to intercept ships coming in at night. There was bright moonlight at first, but around midnight the sky grew overcast so he gave up for the night and bottomed in twenty-one fathoms.

He came up at dawn and at 8.30 sighted smoke on his starboard beam. He altered course to intercept and just after nine o'clock identified her as a big armoured cruiser heading west. Goodhart proceeded at seven and a half knots. At 9.22 the enemy had closed to within a mile and a half. Goodhart eased to five knots to lessen his periscope feather and wake.

(*Top*) Kapitänleutnant Lothar von Arnaud de la Perrière's *U35* arriving at Carthagena 1916. (*Centre*) A torpedo from *U35* sinks another merchantman in the Mediterranean 1916.

British B Class submarines at Venice 1915.

German submarine *Deutschland* meets another U-cruiser in the Atlantic.

Small UB type submarine 1917.

The sun had come out and it was ideal for an attack from the southward. The torpedo officer reported, 'All tubes ready.'

The big cruiser came on, zigzagging very slightly, if at all. She was the *Prinz Adalbert*, Horton's old target, now repaired.

Goodhart stared into the periscope's eye. She was making about fifteen knots. Her port side destroyer escort passed a hundred yards ahead of them. Now she was clear. Range 1,300 yards, Goodhart aimed for the cruiser's forebridge.

'Fire one!' The boat dipped as the torpedo left the bow tube.

'Hard-a-starboard!' The sub swung to starboard to bring her port beam tube to bear. Then Goodhart saw the vivid flash of an explosion along the enemy's waterline at the point of aim, and felt the huge concussion as the cruiser's fore magazine went up and she was hidden in a great column of thick grey smoke.

Big fragments of debris were falling in their wake and all round them so Goodhart took *E8* down to fifty feet. It was just as well that his first torpedo had dealt such a fatal blow, as the shock of the explosion had filled the buoyancy chamber of *E8*'s port beam torpedo. Ten minutes later he came up to periscope depth. The cruiser had disappeared. Only the two destroyers were left, forlornly picking up the few survivors from a sunlit sea.

E19 went back on the Swedish coast patrol. At three o'clock she surfaced in fine weather some twelve miles off Karishamn on the south coast of Sweden to investigate a steamer about six miles away steering south-west. Cromie dived and closed for a better look, then surfaced and hoisted 'Stop instantly' in International Code. The ship took no notice of the flags and ignored a shot across her bows from the deck gun, so Cromie switched fire to her upperworks and hit the foot of her funnel. That stopped her. Cromie went alongside and discovered her to be the *Suomi* of Hamburg with wood from Nydal.

Her captain could raise no objections to the destruction of his ship as they were now a good twenty miles from the nearest land, but Cromie could guess at his feelings as he watched impassively while the boarding party set fire to his cargo and went below to fix a charge under the main inlet. Then Cromie started main engines and had just turned to head south when the charge in the *Suomi* blew. When last seen she was down by the stern, with a heavy list to port and burning fiercely fore and aft.

The weather turned bad, and heavy seas once again prevented boarding, but at 1.20 p.m. on 7th November Cromie, tired of lying off Trelleborg watching the various ferries plying between the Swedish and Danish shores, sighted a two-funnelled light cruiser and a destroyer four miles to the north-east of him steering south-east at about fifteen knots.

Cromie dived to attack, identifying her as an *Arkona* class ship. She was in fact the *Undine* of that class, 2,645 tons, top speed 21.5 knots, ten 4·1-inch guns, two torpedo tubes, a fast, useful vessel built in 1903.

At 1.45 Cromie fired his starboard beam tube from a range of 1,100 yards, and hit her forward on the starboard side. She swung round in a wide circle and stopped, on fire and sinking by the head, with her ship's company assembling on the poop.

Cromie manoeuvred to avoid the destroyer, passed under the cruiser's stern, got the sub into a good position, and ten minutes after the first shot fired a second torpedo, from 1,200 yards, aimed at a point below her mainmast.

There was an explosion just aft of the mainmast, then a second and much heavier blast as *Undine*'s after magazine blew up. Several large chunks of smoking debris shot out some two hundred yards towards *E19*, and Cromie had to dive to avoid the destroyer, which had opened a heavy fire on his periscope. He raised the periscope three minutes later, and there was no sign of the cruiser.

CHAPTER SIX

'Worth an Army Corps'

The German drive on Paris had been stopped, but the soldiers were stuck in the mud. Both sides were stalemated in a line of trenches and fortifications stretching from the flooded coast of Belgium right across France to the Swiss border.

It was almost as bad in the North Sea. After Dogger Bank the German High Seas Fleet was too scared of Jellicoe's mighty battlewagons to come out, and the Grand Fleet was frightened of U-boats.

In February 1915 the Allies began a move, sponsored by Winston Churchill, to break the deadlock on land by capturing the Straits of the Dardanelles, between the Trojan shores of Turkey and the Gallipoli peninsula in her European territories. This was the vital prelude to a drive across the Sea of Marmara on Constantinople and the eliminating of Turkey from the war.

The Russians, driven back by the German army in the west, had asked urgently for help against Turkey, but the French Government and the British War Office did not like the idea of the Dardanelles assault any more than they had fancied Lord Fisher's schemes for seizing German islands in the North Sea or attacking in strength in the Baltic, which had been rejected as too risky. With their iron and coal areas in Alsace and Lorraine already in German hands, the French did not want to divert troops from the Western Front. Churchill argued that the Straits could be forced and Turkey beaten by naval strength, with a minimum back-up of troops.

The scheme offered some glittering prizes. The Russians could be reinforced through the Crimea and supplied with the rifles, guns and munitions that they seriously lacked in exchange for badly needed wheat, oil and wood; British intelligence could get closer to German liaison with the Bolsheviks. All the Turkish armies in Armenia, Syria, Palestine and the Sinai Desert would be cut off from their depots in European Turkey. Italy, Greece and Roumania could well join the Allies. Bulgaria, surrounded by enemies, might think twice about joining the Central Powers. The fall of Constantinople, legendary Byzantium, would ring round the Moslem world and bring peace with Turkey through revolution . . .

It was too tempting. The Dardanelles campaign began in February with a bombardment of the outer forts defending the Straits by an Allied fleet built round fifteen capital ships, including the new battleship HMS *Queen Elizabeth* and the Dreadnought battlecruisers HMS *Inflexible* and *Indefatigable*..

On 8th March the main thrust into the Straits was launched. All seemed to be going well at first. A shower of shells raised a lot of dust over the forts and batteries lining the shores. Then the big ships ran into a new minefield, not marked on their charts. The French battleship *Bouvet* and the British *Irresistible* and *Ocean* were sunk, the *Inflexible* badly damaged.

This early reverse was enough for the British C-in-C, Vice-Admiral de Robeck, who refused to commit his ships to the Straits again. It was decided to land an army on the Gallipoli Peninsula after all.

It would then be important to cut the Turk's best supply lines across the Sea of Marmara to Gallipoli from Constantinople and from Panderma on the Asiatic shore. If surface ships could not do it, Commodore Roger Keyes thought that his submarines might.

A British submarine had already forced her way into the dangerous waters of the Dardanelles, which opened into the Marmara. Ever since the German battlecruiser *Goeben* and the cruiser *Breslau* had eluded Allied forces in the Mediterranean and sailed through the Dardanelles and the Marmara to Constantinople, an Allied naval force had patrolled the entrance to the Dardanelles. When Turkey, controlled by a pro-German clique, joined the Central Powers on 13th October 1914, this squadron included three British and four French submarines. They were all obsolete coastal types. The British *B9*, *B10* and *B11*, with their fickle petrol motors and low speeds, had made a voyage from England to the Eastern Mediterranean that was in itself a small epic.

Lieutenant-Commander G.H. Pownall RN, CO of the Allied submarine flotilla, wanted to do more than just patrol up and down between Cape Helles on the European side of the Straits and Kum Kale on the Asiatic shore. Already Lieutenant Norman Holbrook in *B11* had penetrated Turkish waters and chased two gunboats. The next step for a submarine was to go through the Straits and get at the Turkish warships lying inside under the protection of gun batteries lining the shores.

The possibility of an attack on the *Goeben* in Constantinople had been discussed, but none of the old submarines had the submerged endurance to make the thirty-five miles of mines, patrol boats and strong, treacherous currents to reach and cross the Sea of Marmara. However, in early December permission was given for one boat to try to go through as far as Chanak, about a third of the way up the Straits, where enemy warships lay. Holbrook's *B11*, which had a new battery, was selected as having the best chance of success.

One of the worst natural obstacles was a strong 2½-knot current which would be running against a submarine trying to penetrate the Straits. *B11*, even with her new battery, would be hard put to breast it, and there was fresh water at ten fathoms, liable to upset the trim. Five lines of mines were known to have been laid across the Straits.

While preparations for the attack were in progress in November, the latest information on the positions of the mines was received from Mr Palmer, late British Vice-Consul at Chanak, updating the plan sent by the Admiralty; special guards were fitted to *B11* by the staff of HMS *Blenheim*, under the direction of Engineer Commander William Wilson and Carpenter William Cooke.

A large nosepiece, consisting of a 3/8-inch steel plate dished to fit the shape of the bow and secured by bolts, was fitted to prevent vertical wires or ropes from catching on the bows of the boat or on the torpedo tube caps when they were open for firing, and the covers of the bow caps were removed to stop them touching the nose piece when fully open. A 1¼-inch diameter stay was fitted from the tip of the forward overhang of the plate down to a point in the stem below the centre line to ward off horizontal wires. *Blenheim*'s divers went down and made the holes for securing bolts. A thick curved rod was bolted to each of the permanent forward hydroplane guards to prevent vertical wires getting between the guards and the hydroplanes. Guards were also fitted round the sections of exhaust pipes protruding from the submarine's sides, and round the spindles bearing the rods which moved the after hydroplanes. Finally a jumping wire was fitted over the whole length of the boat from a mooring eye plate on the bows to a bolt screwed into the head of the vertical rudder. The jumping pole raising the wire above the conning tower was made by the Dockyard at Malta and secured in the exhaust ventilating pipe on the conning tower.

After fitting, trials were carried out using a mine sinker suspended on a wire from *Blenheim*'s main derrick. *B11* was driven at full speed against this wire, which was thrown clear by the special guards.

It was at first intended that Pownall himself, an experienced submarine officer, should command the operation, but it was then thought fairer to allow Holbrook to command his own boat.

B11 moved off early on the morning of 13th December. The extra guards fitted round the hydroplanes fouled her mooring ropes when she was getting under way, but at 4.15 a.m. *B11* was lying submerged three miles west-nor'-west of Cape Hellas. It was still dark, and searchlights on both headlands swept the opening to the Straits. As the first red rays of sunrise came up over the Marmara the searchlights were switched off. Holbrook surfaced, and *B11* crept to within a mile of Cape Helles, trimmed down and at 5.22 a.m. started diving.

Holbrook steered for Seddul Bahr on the southern headland, rounded it close and proceeded at two knots up the Straits, keeping at sixty feet, rising about every forty-five minutes and putting the periscope up for a few seconds to check their course, unwilling to show a feather to the gunners ashore. The old sub's compass was on deck and had to be viewed from below through a series of lenses, which made conning difficult.

They crawled on, and at 9.40 a.m. were clear of all minefields. Holbrook came up to see if there was any trade. The periscope broke surface, and he did a sweep. Anchored off their starboard beam was 'a large two-funnelled vessel painted grey', the battleship *Messudieh*. He could see the Turkish ensign flying aft.

The current was strong against them. Petrol vapour filled the boat as she went to full power and manoeuvred to swing her bows round and line up her two tubes on the enemy.

At eight hundred yards, about one point abaft the enemy's port beam, Holbrook ordered, 'Flood tubes.' The water gurgled in.

'Stand by . . . ' The battleship filled his eye . . . 'Fire both!' It was 9.53.

Just as he spoke a sudden eddy swung the boat sideways, the tip of the periscope dipped below the surface, and they were blind. But they did not need eyes as they heard the thud of a hit. When the periscope was clear again Holbrook found the enemy ship now on his starboard beam, issuing a cloud of black smoke. Her guns opened up on him, then a geyser of spray obscured his vision as a shell from one of her guns burst near the submarine, and *B11* dipped again. When he came up this time he had the enemy on his port bow. She was settling by the stern and her guns were silent. Ten minutes after being hit, *Messudieh* capsized and went down, trapping many of her crew inside the hull, though the water was shallow, and most of them were rescued later through holes cut in the bottom plates.

The shore batteries were now firing on *B11* and getting uncomfortably close. First Lieutenant Sydney Winn reported that the helmsman could not read the compass, the lens having clouded over. Holbrook took a careful look round and concluded that they were in Sari Siglar bay, just south of Chanak. The current was carrying them westward towards shoal water. There was a bump as the sub bottomed at thirty-eight feet then stopped, stuck on a mud bank. Holbrook saw sunlight coming in through the glass ports in the base of the conning tower. In a matter of minutes the gunners ashore sighted the conning tower sticking up clear of the water, and shells began to fall around it. Holbrook ordered the helm put hard-a-port and went to full speed.

The air in the boat was foul by now and the batteries almost flat. The engine room staff flogged them to the limit. *B11*'s single screw churned the mud, her old hull groaned and creaked as she strained to get off. Finally, with a great grinding and scraping she made it. Struggling against the current she headed slowly towards the entrance to the Straits. The water was still shallow. She could not get below, and bottomed several times more. With shells pitching near, Holbrook climbed up into the exposed conning tower and in the poor light provided by the thick glass scuttles navigated the sub through the shallows. They reached deeper water. Holbook took her up and steadied her as best he could on what looked like the high ground on the European shore.

He held on in this direction until he could see a clear horizon on his port beam, then altered to port and headed for it at sixty feet, coming up to periscope depth at intervals to correct the course. At 2.10 p.m., about two miles outside the entrance, he blew main ballast and surfaced. He opened the upper hatch. Wonderful fresh air flooded the boat, but it took ten minutes for a match to burn down below and the petrol motor to get enough oxygen to work up to any speed.

Keyes met his submarine commanders on 14th April 1915, ten days before the toops were due to land on the Gallipoli beaches, to discuss the chances of getting a submarine right through the Dardanelles and into the Marmara to attack Turkish shipping.

The difficulties were great. On the surface, patrol boats and shore batteries covered the Straits. Submerged, there were the currents, averaging 1½ knots and running at 4½ knots in the Narrows. *B11* had sampled all these hazards. More recently the French *Saphir*, a more modern boat with a bigger battery, had tried to reach the Marmara. She had got safely under the Kephez minefield but the currents had swept her ashore near Chanak and she had been lost with half her crew. Clearly the old B-boats could not make it, and it was doubtful whether the batteries of Keyes' three newer E-boats would allow them to go for some thirty miles under water without recharging.

The meeting went through all the reasons why the attempt should *not* be made; then Theo Brodie of *E15* said quietly, 'I think it can be done.'

'Then you shall do it,' said Keyes. 'It's got to be tried.'

E15 left Tenedos on the night of 16th/17th April. His orders from Vice-Admiral Eastern Mediterranean Squadron directed him, if not discovered in the Straits, to attack shipping off the port of Gallipoli, at the far end of the Dardanelles, or, if sighted en route, to look for targets off Chanak first.

Submerging off Cape Helles, he got as far as Kephez on the European shore, then the strong current there caught *E15* and swept her on to a sloping sandbank and up to the surface just below Kephez Point, right under the guns of the fort. She could not back off in time. A shell hit her, killed Brodie and opened her battery to sea. Chlorine gas filled the boat and drove her crew up and into the hands of the Turks.

The heavy firing at dawn was heard outside the Straits, and it was clear that something had gone wrong. Theo Brodie's brother, Lieutenant-Commander C.G. Brodie, flew over the Straits in a seaplane and returned about ten o'clock to report that he had seen *E15* ashore in Kephez Bay, with a Turkish torpedo-boat alongside her.

At Brodie's suggestion Lieutenant MacArthur was sent up the Straits in *B6* that morning to try to torpedo *E15* before the Turks could make use of her.

MacArthur entered the Straits at 2 p.m. Near Kephez he came under heavy fire and his compass failed. He could not see *E15* but fired one torpedo at a range of 1,500 yards at a vessel which fitted Brodie's description of the boat which he had seen alongside *E15*. MacArthur thought he had scored a hit, and observers in the battleship *Vengeance* off the entrance to the Straits saw a tug alongside the submarine turn over and sink.

E15 herself remained intact. Bombs dropped near her by aircraft during the day merely succeeded in driving away another boat which was working near her, and that night two destroyers were sent up to find her and render her useless. *Scorpion* got to within half a mile of her supposed position but could see nothing of her. The destroyers then met a heavy fire and returned.

At dawn next day, 18th April, Holbrook VC in *B11* was sent in, but was thwarted by thick mist near Kephez and also came under heavy fire from the batteries which had stopped *E15*. Just after one o'clock in the afternoon the battleships *Triumph* and *Majestic*, with destroyers sweeping ahead, started ponderously up the Straits to blow *E15* to bits with their big guns. But the light was bad and even with an aircraft spotting for them they could not hit her and retired ignominiously.

After the failure of the biggest ships, an attempt was made with the smallest. Torpedo launching gear was fitted to two picket boats, one from *Triumph*, one from *Majestic*, to go up on the night of the 18th/19th.

Lieutenant-Commander (Torpedoes) Eric Robinson of *Vengeance* was to command the sortie, in *Triumph*'s boat. Torpedo Lieutenant Claud Godwin of *Majestic* joined him aboard *Triumph* at 8.30 p.m. to work out the plan of attack.

The two boats would stay together to a point just below the mouth of the Suandere River, keeping towards the centre of the stream to try and avoid the searchlights to the north as long as possible. Then the *Triumph*'s boat, followed after a short interval by the *Majestic*'s, would make a dash for the *E15*, hoping that at least one boat would get past searchlights, guns and guard boats, and torpedo the submarine.

Robinson passed the Suandere at 11.45 p.m. and was at once fired on. The three searchlights above him became eight, three more on the north shore, two on the south, making the night as clear as day. With the beams blinding them they reached about the right spot and Robinson tried a shot from three hundred yards, which missed by thirty yards to the left.

Godwin in *Majestic*'s boat about four hundred yards behind him caught a lucky glimpse of the submarine as a searchlight flicked over it, fired both torpedoes and had a clear view of the second one hitting the *E15* and exploding just forward of her conning tower, just as a shell hit the picket boat aft. She began to settle. *Triumph*'s boat immediately made for her to take off her twelve men. Armourer Tom Hooper had been badly wounded, and it took four complete circles of the stricken boat, now with only about six inches of freeboard, before he could be taken off, as she was swinging wildly this way and that.

Shells were dropping round them all the time and shrapnel bursting overhead, but they were miraculously not hit, though poor Hooper died on the way to rendezvous with torpedo-boats in Morto Bay.

Brodie joined *B6* next morning to make a recce. Arriving off Helles at 4 a.m. they were told by the destroyer *Wolverine* that the picket boat attack had been successful, but carried on into the Straits to check at close quarters. They proceeded on the gas engine until 5 a.m., dived and were off Suandere at 7 a.m. when they went aground three hundred yards from the shore under the Dardanus fort. *B6*'s conning tower and superstructure were above water for nearly four minutes but the gunners above them were asleep and did not open fire until the sub had slid off into deeper water. While they were stuck there MacArthur was able to take a good look at *E15,* just over a hundred yards away. Brodie also looked at her through the conning tower on the bottom, and Brodie, who had such a keen personal interest in seeing the job done, was able to report *E15* to have been 'rendered completely useless'.

Unlike his Admiral, Roger Keyes was not the man to be put off by initial failure. Stoker's Australian *AE2*, which was finishing off repairs at Malta and was not worked up to proper efficiency, was hurriedly summoned, and on 24th April, one week after the destruction of *E15*, entered the lower Straits. They were lucky enough to get all the way to the Suandere on their gas engine before daybreak, then submerged to get under the Kephez minefield. They went through with the mooring wires scraping their hull, then Stoker rose to periscope depth to check her position, estimated that he was in Sari Siglar Bay, sighted a Turkish gunboat off Chanak and torpedoed her.

Now beginner's luck began to turn. *AE2*'s gyro compass broke down, the currents caught her and she went aground on Chanak Point. She wrenched herself off there, then got stuck again further up on the European side. Each time she was seen from the shore, patrol boats were called in, and as Stoker rounded Nagara Point he was forced to dive. With his battery almost out of juice he lay on the bottom until nightfall. He came up, found his way ahead clear, and started the last fifteen miles on his gas engine, recharging as he went. Just before midnight Keyes received the signal, 'Am in the Sea of Marmara'.

On the following morning, 25th April, French, British and Commonwealth troops landed on the Aegean beaches of the Gallipoli peninsula, and met heavy fire from the Turkish batteries on the heights above them.

That same day Otto Hersing's *U21* left Wilhelmshaven. The Turks had asked for U-boats to attack the Allied armada in the Eastern Mediterranean, and this crack boat had been chosen to make the long voyage out to the Golden Horn from Germany. The former Hamburg-Amerika liner *Marzala* was detailed off to refuel and provision *U21* off the coast of Spain.

Meanwhile, in the blue Homeric waters which were Hersing's destination, Commodore Keyes' remaining E-boat, Edward Boyle's *E14*, was about to make her dash for the Marmara.

So far the worst enemy had been the sea, the vicious currents in the deeps, which had tossed *B11* about like a piece of flotsam and swept *E15* to destruction. But the surface currents were known, marked on their charts and described in sailing directions. *E14* would try to make it on the surface. If she went at night she had a chance of dodging the enemy's destroyers and patrol boats. Better the devils you knew . . .

Boyle left Tenedos on 27th April at 1.40 a.m. At 3 a.m. Seddul Bahr was abeam to starboard, and *E14* headed up the Dardanelles on her gas engine.

At exactly four o'clock the Suandere fort caught them with a searchlight and opened fire. With shells dropping ahead Boyle dived. It was time to go down anyway, to get under the Kephez minefield. They crept forward slowly on their battery at ninety feet and came up for a look-see about a mile south of Killid Bahr. They surfaced then and at 5.15 a.m. passed Chanak on the Asiatic shore, with all the forts there firing on them, kicking up white water all around the narrow conning tower and for'ard casing, just awash. Boyle went to periscope depth.

The Narrows were full of small ships and steamboats on patrol as well. Boyle, the sharp opportunist, could not pass up the chance, even in the midst of these many dangers. At 1,600 yards a small vessel, which he coolly identified as a *Berki-Satvet* class gunboat, swam across his sights. He fired one torpedo at her and just had time to see a big spout of water high as the mast shoot up from her quarter; then he had to dip the scope again as some men in a small steamboat were leaning over trying to catch hold of it.

At 6.30 a.m. *E14* rounded Nagara Point to starboard. From there it was twenty more miles to the Marmara. It was covered with patrol boats all the way, as Boyle found out whenever he came up to take a look. Then at 10.15 Gallipoli town was abeam to port, they passed through Gallipoli Strait, and were in the Sea of Marmara.

There were several sailing and small steam craft about and some torpedo boats, and they had to continue diving until 3.40 in the afternoon. They started to ventilate and recharge, but in twenty minutes had to go down again.

It was the same next day, the 28th. From 9 a.m. until half-past seven in the evening they were kept down, then were able to come up and breathe north-east of Marmara Island. Just about midnight a destroyer was sighted ahead. They dived and stayed under until 5 a.m. next morning, then surfaced and charged batteries until eight o'clock, when the motors were so hot, having been running almost continuously for over fifty hours, that they had to stop.

On the morning of 29th April *E14* was patrolling the Gallipoli Strait, the area where there was a concentration of shipping using the port. Four torpedo-boat-destroyers emerged out of the smoke over the harbour at 12.30 but were going too fast for Boyle to get a shot in. At 1.15 there was more smoke, this time from three destroyers convoying two troopships. The sea was glassy calm and the escorts sighted Boyle's periscope as he manoeuvred. Boyle ignored them as they came foaming towards him, firing on him with their forward guns, and concentrated on one of the transports, about a mile and a quarter off. He fired one torpedo, then dived almost under the forefoot of the leading destroyer, taking the boat down steeply as one periscope had been damaged by a near-miss the day before and he dared not risk losing the other one.

He heard a distant thud, and when he came up half an hour later saw two of the destroyers and one transport on their original course, but the remaining trooper, escorted by one destroyer, heading for the shore with a thick column of yellow smoke pouring from her.

At five o'clock he sighted *AE2*. The two submarines joined up and the two skippers compared notes. Stoker shouted, 'I bagged one gunboat. Had rotten luck with torpedoes. I've only got one left now.' They arranged to meet next day.

Boyle sighted *AE2* next morning and was about to signal her when a destroyer appeared and he had to dive. He never saw *AE2* again, and could not find out why she did not make the rendezvous, as he could not get through to HQ on his wireless. In fact, just after he had lost her that day the Australian boat had made an attack on a TBD. Getting into position for a shot, she had hit a layer of denser water, been bounced to the surface, dived, porpoised again almost under the enemy's bows and, holed in several places by shell-fire, had been abandoned by her ship's company.

Meanwhile Boyle was dodging patrol boats, which kept *E14* down most of the day, with one diversion when he rose and stopped a tug towing three *dhows*, which he thought might contain ammunition, towards Gallipoli. The boats were empty, but he scared them into turning back to Constantinople.

On May Day he managed to get three hours undisturbed on the surface in the morning, then was forced down. By this time he was getting very frustrated at the continual bobbing up and down. He said to First Lieutenant Edward Stanley, 'I'm going to get one of those patrol boats. They're always firing at *me*.' At 10.45 a.m. he fired at and sank a small gunboat which looked as if she had been fitted as a minelayer. This tiddler went down in less than a minute, and shortly afterwards a bigger gunboat came up. He fired at her but the torpedo did not run straight. The gunboat sighted *E14*'s periscope feather in the smooth water, opened fire and made to ram at full speed. Boyle dodged the attack and fired again, but the track was easily seen and avoided on the glassy sea. In the afternoon he saw a tramp steamer heading west, and not for the first time wished for a deck gun. In the end they managed to turn her back by firing rifle shots across her bows.

On 10th May he torpedoed and sank another transport, the big *Guj Djemal* carrying 6,000 troops and a battery of field guns to Gallipoli. After that Boyle had only one defective torpedo left, suffering from a broken air pipe which could not be repaired on board, but he was ordered to hang on in the Marmara because of the intimidating effect of his mere presence on Turkish supply movements. The Turks were seeing periscopes everywhere, and hardly anything other than a *dhow* or a patrol boat was moving. To improve the bluff, Boyle's engine-room staff made a realistic looking mock-up of a deck gun from a piece of piping, an oil drum and some canvas. When they were a mile north of Nagara on their way out through the Straits they had to leave the battleship *Torgud Reis* four hundred yards to port, with nothing bigger than a rifle bullet to fire at her.

Two days later, at 2.45 a.m. on 19th May, Martin Nasmith's *E11* entered the Straits to replace Boyle. As he passed the *Torgud Reis* she and her destroyers opened fire on his periscope. However, he found the Marmara easier to reach than he had the Baltic, and passed Gallipoli at 9.30 p.m.

Nasmith made at once for the eastern end of the Marmara and

Constantinople. On 21st May he sent his First Lieutenant D'Oyly-Hughes to board a small sailing vessel. There was no contraband on board but Nasmith moved the submarine alongside her, trimmed well down, and lashed the boat to his conning tower. That way he was invisible from one side, and able to make good speed with his gas engine for some time.

Early on the morning of the 23rd he arrived off Constantinople and at 5.50 a.m. torpedoed and sank a torpedo gunboat anchored off the port. As she sank, the Turk fired on *E11* with her 6-pounder, and the first round hit and damaged the submarine's foremost periscope. Nasmith took them away from the scene and sent hands to bathe in the warm waters of the Marmara.

At 10.30 next morning he sighted a small steamer heading west, surfaced on her quarter and signalled her to stop. The ship took no notice, then a few rifle shots brought her to her senses, she stopped, and Nasmith ordered her crew to abandon ship. There was a panic to leave the ship, and the first two boats away capsized. A man on the upper deck shouted across to Nasmith.

'Sir, I am Silas Q. Swing of the *Chicago Sun*. Pleased to make your acquaintance. This ship was heading for Chanak with Turkish marines. I'm not sure if she carries any stores.'

Nasmith thanked Mr Swing for his information, then the American, with two or three of the Turkish crew, hoisted out a third boat under Nasmith's direction, righted the other two boats and rescued the swimmers. *E11* went alongside, and D'Oyly-Hughes boarded with a demolition party. He discovered a 6-inch gun lashed across the top of the fore hatch, one large 6-inch mounting and several small 12-pounder pedestals in the fore hold, with the guns themselves underneath them. The after hold was full of 6-inch shell with fifty large cartridge boxes on top marked Krupp. A charge was fixed against the ship's side in the after hold, well tamped with 6-inch shells and cartridges. All hands returned to the submarine and the charge was fired. The ship disintegrated with a tremendous explosion and a big column of flame and smoke.

As the smoke and the shower of debris subsided, Nasmith sighted more smoke to the east. He dived to attack but the ship must have seen them, as she altered course towards the port of Rodosto. *E11* surfaced again and chased her until she secured to pier. Nasmith then dived and, with the submarine bumping along the bottom of the inshore shallows and rifle bullets pinging off the conning tower, closed to torpedo range. At 12.35 p.m. a torpedo left his port bow tube and hit the target squarely amidships. The heavily laden supply ship, her deck piled high with packing cases, burst into flames.

As he left Rodosto Bay Nasmith sighted a paddle steamer with rolls of barbed wire on her decks. She was hailed and slowed down, but when *E11* tried to range alongside her she tried to ram the submarine. Nasmith's superior speed got him out of trouble, and he chased the enemy as she ran for the shore on the north side of the bay, skilfully keeping her stern towards them, to make a shot with a torpedo more difficult, paddles thrashing the water. She reached the beach under steep cliffs, and D'Oyly-Hughes was just preparing to board her when a party of horsemen appeared on the cliffs above and opened a hot rifle fire on the conning tower. *E11* was forced to retreat, scraping her way out into deep water. She then fired a torpedo from a stern tube, but it was a long shot at a small target. The torpedo passed harmlessly along her side and exploded on the beach. Nasmith surfaced at five o'clock, recharged his batteries and headed slowly on the surface towards Constantinople. At 12.30 p.m. on the following day, 25th May, he raised the periscope inside the great harbour, the first enemy to enter it for five hundred years.

Moored alongside the Arsenal he saw a large ship, with a smaller one ahead of her. He fired one torpedo, but it failed to run. He fired another, and just had time to see it heading for the larger ship when *E11* was caught by a fierce cross-tide. He heard two explosions, dived, hit bottom at seventy feet, bounced up to forty feet, then checked any further rise by going astern and flooding internal tanks. The current swung *E11* round in a full circle. Nasmith got her head round to the south, and she bumped her way out of habour, surfacing twenty minutes later well clear of the entrance.

Off Cape Helles on the Gallipoli shore three British battleships lay. Hersing in *U21* got out his fleet book and decided they were all pre-Dreadnoughts of the *Majestic* class. He was not quite right. The old *Majestic* herself, 14,900 tons, built at Portsmouth in 1895, with her four 12-inch and twelve 6-inch guns, was one of the three, but the others were the smaller *Swiftsure*, 11,800 tons, four 10-inch and fourteen 7.5-inch, built in 1904, and the 16,500-tons appropriately named *Agamemnon*, four 12-inch and ten 9.2-inch, of 1904. They were all firing their big guns, at Turkish positions in the hills.

There had been rumours, but nothing confirmed, in the British ships of German submarines in the Mediterranean. One had been reported passing Gibraltar, one off Malta, a third in the Doro Channel. Or were they the same boat?

In any case the ship's company of *Swiftsure* was well trained, and when someone shouted, 'Submarine! Submarine!' her gunners manned their 14-pounders without delay and opened up at the periscope three hundred yards away.

This proof of the presence of enemy submarines off the beaches caused consternation throughout the Allied fleet. The comparatively modern *Agamemnon* immediately received orders to weigh anchor and go back to the fleet anchorage at Mudros, as she was too valuable a ship to risk.

Two hours after their first sighting of the U-boat they heard aboard *Swiftsure* that the submarine had fired a torpedo at the old battleship *Vengeance*, which was cruising between Cape Helles and Anzac. The torpedo missed, passing across her bows. The *Vengeance* then puffed off to the doubtful protection of the nets in Kephalos Bay.

Off Gaba Tepe beach, Hersing's periscope showed another old battlewagon of the *Majestic* vintage, her four 12-inch and numerous 6-inch belching fire and smoke, and the shells kicking up dust in the hills behind the beach, where the Allies were dug in. Her anti-torpedo nets were rigged round her on booms, like skirts, and a swarm of destroyers and patrol boats surrounded her.

In *Swiftsure*, anchored further offshore, they were just finishing lunch in the wardoom, when a young signalman approached the Commander with cap in hand and said, with a most apologetic air for interrupting the great man's lunch, 'Beg pardon, sir, the *Triumph* is sinking.'

They all rushed on deck, where every officer gathered, including Rear-Admiral Stewart Nicholson, and there, sure enough, off the beach lay the unlucky old ship, with a heavy list and obviously stricken to death. Destroyers were rushing to her assistance, almost covering the horizon with dense clouds of back smoke. Fortunately there was a trawler close to the *Triumph* at the time, which was able to take off some of her crew. The battleship hung at an angle of forty-five degrees for about eight minutes, then she turned bottom upwards, floating in this position for some twenty minutes, looking like a whale at rest. The Admiral, the officers and ship's company of *Swiftsure* stood to attention, bareheaded, when *Triumph* made her final plunge beneath the waves in a cloud of smoke and steam.

Hersing hid in the blue Aegean for two days, then returned to the beaches. Those still asleep aboard *Majestic* anchored off Lancashire Landing at dawn on 27th May were awakened by the shock and the dull, heavy explosion of a torpedo about fifteen feet forward of the

U-boats: three units of the Weddigen Submarine Flotilla, photo-graphed beside the *Saar* off Kiel just before World War 2.

Submarine *M2* launching Parnall Peto aircraft 1929.

Grossadmiral Karl Dönitz.

U487 of the 12th Flotilla at s
in World War 2.

shelter deck on the port side. The ship lurched to port, and lay on her side. Soon the sea was crowded with swimmers.

A few minutes later the *Majestic* rolled right over to port, and began to sink in the great swirl of white foam and escaping steam. As she sank stern-first and the water rose over her green barnacled bottom towards the bows a sailor ran the whole length of her keel and sat astride the ram, which remained sticking out above the water, with the wreck stuck on a sandbank. When this man was rescued his first words were, 'I didn't even get my feet wet.' Some men were dragged down by the torpedo nets, but the ship went down in shallow water, with a minimum of suction, and many were rescued by boats from the other ships, so loss of life was reduced to fifty men. Captain Talbot was thrown into the water when the ship sank, but was picked up by a launch. Then, seeing two of his men in danger of drowning, he dived into the sea again and saved them both.

While Hersing was stalking the *Majestic*, Nasmith in *E11* was shaping to attack the big battleship *Barbarossa*, heading at full speed through the Marmara Channel for the Dardanelles. He put up the periscope for a final assessment before firing, and found himself staring up the hawsepipe of an escorting destroyer, her bows filling the glass.

'Dive, dive, dive!' Nasmith shouted, and the keel of the thundering destoyer cleared their jump wire by inches. *E11* headed east, came up in the centre of the Marmara, and the crew gave the boat and themselves a much needed wash in the sun. Nasmith wrote euphemistically in his report, 'We had of late noticed the atmosphere in the boat becoming very oppressive, the reason doubtless being that there was a great deal of dirty linen on board and also the scarcity of fresh water necessitated a limit being placed on the frequency of personal washing.' By 8 p.m. ablutions were over and *E11* headed for the southern shores of the Marmara.

At 4 a.m. on 31st May she dived and crept into the harbour of Panderma, the important port and railhead on the northern coast of Turkey in Asia. By eight o'clock Nasmith had his eye on one of the latest ships of the Rickmers Line, lying in the Roads. Completely unobserved, he was able to take his time and set up the target for a perfect hit amidships on her port side. The whole harbour was thrown into a panic. The stricken ship was towed ashore, listing heavily to port. Not wishing to push his luck, Nasmith left the harbour and headed north-east. Just after nine o'clock on the morning of 2nd June he fired a torpedo at a ship sighted to the east of them. He had found an ammunition ship. When the tinfish hit her on her port side the resulting explosion blew the entire upper deck overboard, and the shattered wreck which remained sank instantly.

There was always a price to pay for success, and *E11*'s hard-working machinery was showing signs of strain. On 5th June they had an earth on the port main motor armature, and the starboard intermediate shaft was found to be cracked. It was time to wrap up the patrol and go home.

At 3.40 a.m. on 7th June *E11* dived and re-entered the Straits. Three hours later she passed Gallipoli at ninety feet. As well as being a brilliant underwater tactician and ship-handler and a good shot, Nasmith was a careful ship's husband. Thriftily, he always had his torpedoes set to float after a run, so that he could recover those that missed, and *E11* still had some shots in her locker. As they glided slowly towards Nagara he searched every anchorage for battleships, especially the *Barbarossa,* which had so nearly been their doom.

He saw a big empty troopship anchored off Moussa Bank, but held on to his precious torpedoes in hopes of sighting battlewagons lower down. They rounded Nagara Point and searched the Straits as far as Chanak. They found nothing interesting, so *E11* turned sixteen points, went upstream again and torpedoed and sank the big transport.

Immediately after rounding Kilid Bahr the submarine lost her trim, and she needed eight tons of water ballast to get her down to safety at seventy feet. An hour later, about three in the afternoon, there was a loud grating noise as if the boat were grounding, which was impossible at this depth. Nasmith went up to twenty feet to investigate, took a look through the periscope and saw the frightening sight of a large mine hanging by its moorings to his port hydroplane, trailing about twenty feet in front of the periscope.

He could not bring the boat up to clear it because of the batteries on shore. Carefully he took the lumbering, yawing submarine out of the Straits at thirty feet, and rose to twenty feet when they were the right side of Kum Kale. He went astern at full speed and ejected water from the after tanks to leave the bow submerged and bring the stern to the surface. The rush of water from the screws and the stern way gathered allowed the mine to drift clear and it fell away ahead of the boat.

In this epic patrol Nasmith had sunk a gunboat, three transports, an ammunition ship and three store ships, all without a deck gun, which was badly needed in these boats. This one submarine had been, in the words of General Sir Ian Hamilton, commanding the troops ashore, 'worth an Army Corps'. Nasmith, who had already been mentioned in despatches for rescuing five airmen from the sea while under attack by a Zeppelin in the Heligoland Bight on Christmas Day 1914, was awarded the Victoria Cross.

CHAPTER SEVEN

'Kamerad!'

With the bloody failure of the Suvla landings in August the imaginative plan to cut Turkey off from Europe finally collapsed, and there was only the long agony of evacuation left for the Allied army.

The British submarines continued their attacks on Turkish supply lines until the end of the year. Nasmith entered the Marmara in *E11* for the last time on 6th November. In a record patrol lasting fifty-two days he sank a destroyer, eleven steamers and thirty-two *dhows*.

By this time Q-ships, the 'mystery ships' which looked like ordinary coasters or small freighters but carried naval crews and concealed guns, had seen action against U-boats. A Q-ship trailed her coat in known U-boat hunting grounds and if stopped sent away a 'panic party' in the boats, with guns' crews left behind ready to unmask the armament, hidden behind false bulwarks or even mock deck-houses, when the U-boat came within range.

An early variation on the more 'orthodox' Q-ship was the trawler towing a submerged C class submarine, with a telephone link. A combination of Lieutenant-Commander H.D. Edward's trawler *Taranaki* and Lieutenant F.H. Taylor's *C24* surprised and sank *U40*, cruising off Aberdeen on 23rd July 1915, after the U-boat had fired a warning shot across *Taranaki's* bows, and the sub/Q unit *Princess Louise* and *C27* sank *U23* a month later off the Orkneys.

U36 stopped a coaster in the same area on 24th July and had just surfaced when her captain saw another small merchant ship approaching. Eager to bag the pair, he turned his attention to the newcomer, which stopped and lowered her boats when the shell whistled across her bows. *U36* closed this vessel, still firing on her. When the U-boat was six hundred yards off, Lieutenant Mark-Wardlaw, commanding the Q-ship *Prince Charles*, unmasked her two 6-pounder and two 3-pounder guns and returned the fire. Her broadside was too heavy for the U-boat, which sank, leaving her captain and fourteen crewmen in the water.

On 19th August 1915 the *U27* surfaced and stopped the British freighter *Nicosian*, allowing her crew to get clear in the boats. They, and particularly the eight American sailors among them, then watched with amazement when another vessel, apparently a merchantman, with the Stars and Stripes flying aft and painted on her side, came up, dropped panels to reveal guns, opened fire on the U-boat and sank her.

The strange ship, having first rescued all the *Nicosian* men, shot all the German sailors in the water, boarded the *Nicosian*, found six more U-boat men who had scrambled aboard her, and left them dead also. The U-boat captain, who had been with them, jumped overboard and was shot.

The *Nicosian*'s American crewmen complained to their State Department about the incident, and it was revealed that the strange intruder was the British Q-ship *Baralong*. In his report her captain claimed that his Royal Marine boarding party had found Germans aboard the *Nicosian* who 'succumbed to the injuries they had received from Lyddite shell'.

This incident increased anti-British feeling in certain sections of American opinion, including businessmen who had had letters or cables delayed, business secrets allegedly filched by British censors, cargoes held up or confiscated by the British blockade, but on the same day as the *Baralong* affair Schneider's *U24* sank the unarmed *Arabic* without warning sixty miles south of Queenstown. Of 181 passengers, some of them American, and a crew of 248, forty-four lives were lost. Isolationists were able to say 'a plague on both your houses'. On the following 24th September the *Baralong* gave a repeat performance of the *Nicosian* savagery, this time against the crew of the *U41*, which had stopped the steamer *Urbino*.

On the night of 19th/20th March the young Oberleutnant Pustkuchen's *UB29* from Zeebrugge torpedoed three neutral ships lying at anchor off the English coast, and on the 24th torpedoed the Folkestone-Dieppe ferry *Sussex* without warning, killing fifty men, women and children, including some American citizens, reporting her as a transport. This followed the cynical destruction of two neutral liners in the Mediterranean by the notorious Max Valentiner's *U38*, which had brought protests from President Wilson. Now Wilson threatened Germany with a break in diplomatic relations and the Kaiser ordered a return to the strict observation of international law with regard to the proper warning and searching of ships and provision for crew safety.

The Cattaro flotilla was ordered not to attack any passenger ships, even those suspected of being armed merchant cruisers. Outstanding in the Mediterranean was Lothar von Arnaud de la Perrière in *U35*, a tough and clever sailor, great-grandson of a French officer who had served Francophile Frederick the Great and settled in Prussia. Before the war von Arnaud had been torpedo officer in the light cruiser *Emden*, later a notorious commerce raider. Then he was aide-de-camp to von Tirpitz, and was on the Staff when war broke out. He asked for Zeppelins but there was no vacancy, so he went to the other extreme and entered the submarine service. He had only taken over his first command in January 1916 but soon led the pack. He was meticulous in observing the rule of law at sea and the dictates of humanity. Merchantmen were always warned, their crews given ample time to man the boats. Yet von Arnaud still managed to run up the highest score of any U-boat commander. In June he sank thirty-nine ships totalling 56,818 tons. In a cruise in August his score was fifty-four ships (90,150 tons) in twenty-four days, an all-time record for a submarine commander.

In January 1917 *Q5*, a former collier, was in Plymouth refitting when her captain, Lieutenant-Commander Gordon Campbell, learned of the new German unrestricted submarine campaign, which the Kaiser, under pressure from the stalemated German Army, had declared. He decided that the only sure way of decoying a U-boat to the surface for his guns to get a lethal hit was actually to encourage the sub to torpedo him.

At 9.45 on the morning of 17th February *Q5* was about sixty miles west-sou'-west of Cape Clear, Southern Ireland, when a lookout shouted, 'Torpedo approaching from starboard side!'

The torpedo had been fired from long range and there was ample time to avoid it, but Campbell, following his plan, did nothing until the very last minute, when he put the helm over to avoid a hit in the engine room and the inevitable loss of life there.

The torpedo exploded against the bulkhead just abaft the engine room with a great crash, knocking men down, including Campbell, and the sea started to flood the ship.

The panic party rushed for the boats, and the U-boat closed them, still submerged. Her periscope's optic raked the disabled ship from about ten yards, so close that Campbell, crouched low on the bridge, could see her whole hull under water, and was tempted to open fire, but held off for a better target.

The ship was getting lower in the water all the time, and the after gun was practically awash. The guns' crews fingered firing lanyards, the wireless operator's finger twitched towards the morse key.

Would he just move off, leaving them with nothing but wet feet and red faces? The U-boat passed up their starboard side and crossed their bows, and her conning tower hatch opened. As she came abreast of the ship Campbell saw her captain climbing out of the conning tower hatch, and gave the order to open fire.

Flaps on *Q5*'s sides opened, her dummy wheelhouse collapsed and a hen coop spat Maxim gun fire. Her first shell, from a 6-pounder, hit the sub's conning tower, and the captain dropped down the hatch.

It was point-blank range. Concentrated fire from *Q5*'s five 12-pounders, two 6-pounders and one Maxim gun smashed the U-boat's conning tower, and *U33* sank. *Q5*'s boats were only able to rescue one officer and one seaman. *Q5* herself was beached at Berehaven.

A few U-boats were being sunk by determined actions like this one, but the loss of ships in February rose to 230, totalling 464,599 tons, an increase of fifty per cent over the January losses. Nearly three hundred neutrals were refusing to sail from British ports or to Britain from abroad, and others were ignoring orders to put into British ports for examination.

In February *U54* sank five ships without warning south-west of Ireland, but one ship attacked turned out to be the *Q15*, which damaged the sub with two of the new depth-charges. A thousand of these, with a hydrostatic pistol and a charge of TNT, had been ordered in August 1915, but delivery was very slow, as was the development of hydrophones for detecting the sound of a U-boat's screws.

Sir Maurice Hankey, Secretary of the Committee of Imperial Defence, advocated a 'system of scientifically organised convoys', with all the means of attack on submarines concentrated round them, but the Admiralty snubbed this idea, believing that a convoy was an invitation to a massacre. Merchant captains thought that they would be unable to manoeuvre together or keep station, especially at night. The success of an experimental convoy of colliers sailed to Brest from Cornwall and another of nine ships safely escorted from Berwick to Bergen was ignored.

On 1st March a telegram from German Foreign Minister Zimmermann urging Mexico to bring Japan into an attack on the USA, should she declare war on Germany, was intercepted by US Intelligence. This proved to be the last straw for the American public, and events moved steadily towards war.

U-boat sinkings in March exceeded half a million tons (507,001) for the first time, with only two U-boats being destroyed, one by the British submarine *G13*, the other by the Q-ship *Privet*. Nine new U-boats were commissioned. Germany also began a special campaign against hospital ships, many of which, they claimed, carried troops and munitions. On 20th March the *Asturias*, sailing with all her Red Cross symbols clearly illuminated, was torpedoed and sunk without warning, and on the 30th the *Gloucester Castle*.

These actions created more international ill-will against Germany, and on 6th April the USA finally declared war on her. Rear Admiral W.S. Sims, President of the US Naval War College, was already on his way to England. On 9th April he met Admiral Jellicoe, now First Sea Lord, in the Admiralty and was shown the grim figures for sinkings by U-boat for February and March, with a forecast of an increase to a million tons in April. His report to the American Secretary of the Navy described the U-boat campaign as 'the real issue of the war' which the Allies 'are not now effectively meeting'.

April losses did climb steeply, and many men, cargoes and fine ships were destroyed. Four days before the US declaration of war the American *Aztec*, 3,727 tons, was sunk off Cape Finisterre. On the 8th Kapitänleutnant Werner in *U55* sank the British steamer *Torrington* 150 miles south-west of the Scillies. Her crew got clear in two boats but all the men in one boat, including the master, were removed on to the upper deck of the submarine. The master was then taken below, the boat smashed open, and the other twenty men left to drown as the U-boat submerged. None of the men from the other boat were ever seen again either, and the master recognised some of their gear in the hands of German sailors. It was a new note of savagery in the U-boat war, and Werner went on to murder the crew of the 3,066-ton *Toro* on the 12th two hundred miles west-nor'-west of Ushant.

A convoy of sixteen merchantmen sailed from Gibraltar on 10th May, weakly escorted by three Q-ships and three armed yachts but met a hundred miles from Ireland by six destroyers and brought in without any trouble.

A flying boat arrived over the ships when the destroyers joined and carried out searches ahead until the convoy got in. By this time in the war aircraft were patrolling out to sea from air stations on the east, Channel and west coasts of England and the north of Scotland. These machines were slow, frail and unreliable, and very much at the mercy of the wind and weather, but the best of the flying boats could cover an area of some 4,000 square miles and sometimes reach as far as seventy-five miles from the coast. They carried only tiny bombs which had to score a direct hit to do any damage to a U-boat, but flying boats bombed eight U-boats in April, some valuable sighting reports were passed to ships, and there were non-rigid airships, or 'blimps', which could remain airborne for many hours and operate at night. The aeroplane was now something else a U-boat had to watch out for.

Ten ships left Hampton Roads, Virginia, in convoy on 24th May, and managed to keep station and even zigzag together through the danger zone to arrive in Britain almost intact. Only one ship, a straggler, was sunk, and with the small, six-ship convoys which crossed the North Sea between Britain and Scandinavia it was found that only when ships became detached or scattered were they in serious danger from U-boats.

Sixty ships in four convoys reached Britain from the USA in June with only one ship lost, and the new Convoy Committee of the Admiralty had plans for convoys from Gibraltar, West Africa, Hampton Roads and New York, though these were still only paper fleets, and homeward-bound ships from Gibraltar and the South Atlantic as well as all outward-bound vessels were to continue to sail independently.

These ships presented continued rich spoils for the U-boat arm, its losses made good and its operational strength now increased to 150 boats, fifty of them at sea in British waters, some ten or twelve in the Mediterranean. Sinkings, down in May, were up again in June to 631,895 tons.

U-boats were wise to Q-ships, and torpedoes were once again the main weapons, against which all merchant hulls were equally vulnerable. Anti-submarine submarines had taken their place, and there were hunting flotillas of patrol boats armed with hydrophones and depth-charge throwers, though the hunter subs were slow and their torpedoes unreliable, and the hydrophone was hardly a precision instrument, making the tracking of a submerged submarine difficult.

In August the veteran Otto Hersing, sinker of the *Triumph* and *Majestic* in the Mediterranean, patrolling off south-west Ireland in his old *U21*, attacked a convoy of fifteen ships escorted by no fewer than twenty-four destroyers. He claimed two hits but was hunted for five hours and badly shaken by a massive barrage of depth-charges.

Escort in this sort of strength was rare, and the August figures for merchant ship sinkings gave great cause for alarm to the Allies.

They meant nothing, however, to the hungry German public, promised by the admirals that the U-boats would have Britain begging for peace in six months from the start of the new unrestricted campaign. Six months had gone by. The offensive had brought the Americans into the war, but there were no peace overtures from Britain. There *was* a peace movement in the Reichstag, reflecting the increasing war weariness of the public, and some of the deputies had been stirring up trouble in the High Seas Fleet, where there were minor mutinies in August.

In September Allied merchant ship losses were down to 315,907 tons sunk. It was still a high figure but a significant drop, particularly as many of the ships lost were small coasters sailing alone. Of the sixty U-boats operating, eleven were lost, but the U-boat arm was still very large and determined, uncontaminated by the poison of unrest in the High Seas Fleet, and construction was still keeping ahead of losses.

Korvettenkapitän Hashagen was on patrol in the Atlantic in October in *U62*, when he sighted the American steamer *Luckenbach*, which was straggling a hundred miles behind her convoy. This was the sort of target U-boat captains now hoped for.

Hashagen fired a shot across the American's bows. She fired back and increased speed, and a running battle developed. *U62*'s 4·1 outranged the merchantman's gun, and Hashagen hit her with a dozen shells, setting her cargo of Southern cotton on fire, but she steamed on, her wireless operator sending out a steady stream of SOS signals, which brought along the *Nicholson*, one of the US destroyers based on Queenstown, Southern Ireland.

The destroyer opened fire, and her second shell hit *U62* in her bow. The sub dived and was kept down but not damaged by depth-charges. When Hashagen came up again he sighted a convoy of twenty merchantmen and ten escorting destroyers. Leading the second column was the big British armed merchant cruiser *Orama*, and *U62* sank her with her last torpedo.

To attack a ship in the middle of a convoy was a difficult and dangerous gambit. Chance and luck had helped Hashagen sink the 12,927-ton former Orient liner. Most sinkings around convoys were from attacks on stragglers or on ships after they had left or just before then had joined their protected ocean caravans, Hashagen sank the Cunarder *Ausonia* that way, six hundred miles out in the Atlantic, as she steered her lone course west after her convoy had been dispersed.

In the Baltic organised convoys and stronger anti-submarine measures by surface warships had made the British submarine victories of 1914 and 1915 things of the past. Aircraft were playing a bigger and bigger part in fighting the British submarine flotilla there, and both *E1* and *E19* were heavily bombed several times in the spring and summer of 1916. A submarine, with her upper hull painted green and brown and her hydroplanes bright red, could be seen from the air even at a depth of eighty feet. Silver Zeppelins were also to be seen cruising over the shipping lanes.

In November 258,521 tons of Allied shipping were sunk by U-boats, UB-boats or the mines of the UCs; in December 353,083 tons, some of the losses being among ships transporting and supplying the American army in France, which encouraged the German High Command to think that a new offensive in France, coupled with a stepping-up of the U-boat campaign, might yet win them the war. The Treaty of Brest-Litovsk with the Bolsheviks in Russia gave them hope of Ukrainian wheat to make up for the food supplies lost to the British blockade.

In December the U-boat Office was created, to co-ordinate all construction, training and operations in all theatres under one head. U-boats were still sinking ships faster than the Allies could replace them, while their own losses were being exceeded by the number of new boats, and a new programme of over a hundred boats was now begun. In spite of Allied efforts to close the Heligoland Bight and the narrow Straits of Dover, the submarines which could yet win the war for Germany were getting through.

USS *Nicholson*, the destroyer which had saved the *Luckenbach* from Ernst Hashagen's *U62*, was flagship of a division of 'tin cans' on convoy duty in British waters. On 17th November the division lay in Queenstown harbour waiting to escort a US-bound convoy.

Agents of the Irish Republican Army had passed information about this convoy to German Intelligence, and waiting outside Queenstown was Kapitänleutnant Gustav Amberger's *U58*, his special target the big SS *Welshman*.

Just before noon *Nicholson* led the other five cans out of harbour, USS *Fanning* bringing up the rear.

Coxswain 'Eagle Eye' Loomis of *Fanning* saw *U58*'s asparagus. *Fanning* attacked with depth-charges and blew the submarine to the surface.

Amberger climbed out of the conning tower with his hands in the air:

'*Kamerad! Kamerad!*'

Lieutenant Carpender, *Fanning's* CO, rescued thirty-five men from *U58* and got a towline aboard her, but the U-boat was in sinking condition, snapped the wire and disappeared.

To exploit the advantage gained by the growing success of convoy, Jellicoe was replaced as First Sea Lord by Admiral Rosslyn Wester Wemyss, Admiral Bacon at Dover by Rear-Admiral Roger Keyes.

The aggressive Keyes, whose submarines had snatched their own victory out of the general defeat at the Dardanelles, was now a poacher turned gamekeeper. The first thing he did was to flood the narrow Straits with searchlights at night so that U-boats could no longer creep through on the surface to avoid the thick minefields. In April 1918 five of them were destroyed there. The Heligoland Bight was thickly mined, and U-boats going northabout had to detour through the Kiel Canal into the Baltic and out into the North Sea through the Kattegat.

Once out, a new boat with a largely green crew found few easy targets. The hard core of veteran submariners left were spread thinly over the U-boat fleet, and convoys were hard nuts to crack.

In May new tactics were tried. In reports of single attacks on convoys by veterans like Hersing and Hashagen, Michelsen, commander of the U-boat arm, noted that at the first sign of attack several escorts left their stations screening the merchantmen to concentrate on the U-boat, leaving gaps through which other attacks could have been made.

Between 12th and 15th May a force of nine U-boats was assembled west of the Scilly Isles to work together against convoys. It was the first attempt at using a 'wolf pack'. Nine convoys were in or near the pack's patrol zone immediately after it had gathered and if a boat sighting a convoy had continued to shadow it while homing in other boats by wireless, a killing might have been made, but only unco-ordinated attacks were carried out. In a fortnight 293 ships passed through the danger zone and only three were sunk.

Like the British with convoy, the German Navy had a winner but did not realise it. Unlike the British they did not persevere with the idea. The convoy system in the Mediterranean was now halving the losses there, and with the English Channel blocked, half the Flanders U-boats were withdrawn to Germany.

The big 2,000-ton U-cruisers, with their two 5·9-inch guns and tremendous endurance, had great potential, though they were unwieldy under water and had hitherto only been used on the more far-flung trade routes. *U155*, the first to go on active service, had sunk nineteen merchantmen totalling 53,000 tons before returning to Kiel in September 1917 after a patrol lasting over three months. So far, although ten of these boats had been ordered, it had only been possible to have a maximum of two on operations at the same time.

U151, commanded by Korvettenkapitän von Nostitz und Janckendorff, spent a month off the American coast in May-June, in the course of which she laid mines off the US Navy base at Norfolk, Virginia, and Delaware Bay, which sank four ships, cut two trans-Atlantic cables, sank twenty-three small merchantmen, and was nearly sunk by her own armed merchant cruiser, the *Prinz Wilhelm*. Von Nostitz rescued all the crews and passengers from his victims, and even kept some aboard the submarine for a time.

U156, *U140*, *U117* and *U152* were other big boats which operated off the US coast, and forty-two ships of over 1,000 tons each were sunk there by the torpedoes of the U-cruisers, and over fifty smaller craft. Six ships, including the cruiser *San Diego*, sank on their mines, and the battleship *Minnesota* was badly damaged. Von Scheer wanted to use the U-cruiser in conjunction with wolf packs in European water, but was overruled.

There were a few vague attempts at multiple attacks on convoys, following the debacle in May. On 19th July Convoy OLX39, with seven ships, including the 32,234-ton liner *Justicia*, and seven destroyer escorts, left Belfast for the USA.

The *Justicia* had been under construction in a British yard for the Holland-Amerika Line as the *Statendam* when war broke out. The White Star Line then took her over as a troopship, and she was on her way to pick up another contingent of American doughboys for France.

The convoy was sighted at midday by *UB64*, commanded by Oberleutnant von Schrader, and at ten minutes to two, twenty-two miles south of Barra Head, southernmost point of the Outer Hebrides, she hit and stopped the *Justicia* with one torpedo on her port side. Escorts dropped depth-charges uncomfortably close to von Schrader, but at 4.15 p.m. he hit her again. A heavy depth-charge attack left *UB64* trailing oil, but at 7.48 the stubborn skipper scored a third hit.

At seven o'clock on the following morning the *Justicia*, under tow, was sighted by Kapitänleutnant Wutsdorff's *UB124*, which fired two torpedoes from a distance of two and a quarter miles and heard one hit, then went out of control through a mistake in diving drill.

At 8.40 a third U-boat, Kapitänleutnant von Ruckteschell's *U54*, came on the scene. Destroyers chased him, but he managed to open fire with his stern tubes at the slow-crawling liner, and heard an explosion. *Justicia* sank at 12.30.

At four o'clock in the afternoon Wutsdorff blew his tanks. *UB124* shot to the surface, her bow almost vertical, the bilge water swamped the electric motors, which gave off chlorine gas. The destroyers *Marne*, *Milbrook* and *Pigeon* immediately opened fire, and Wutsdorff scuttled his boat, her crew being rescued by *Marne*.

The great German spring offensive on the Western Front had failed, and the Allied armies, stiffened by the Americans, were successfully counter-attacking. The hungry, totally war-weary German public blamed their vaunted U-boats for not stopping the doughboys in mid-Atlantic. Turkey and Austria-Hungary were beaten, Bulgaria was cracking.

Still von Scheer, who had become Chief of Naval Staff in August, looked forward to the launching of the 124 new submarines due for completion in 1919.

Lothar von Arnaud returned to Germany in the spring of 1918 to take command of the U-cruiser *U139*. These boats, which had been operating successfully in small numbers off the US coast, had now been given names as well as numbers. *U139* was called the *Korvettenkapitän Schwieger*, after the man who had sunk the *Lusitania*, the act which finally established the terrible power of the submarine.

U139 was very different from the old *U35*, twice as long (400 feet), displacing 1,930 tons on the surface and 2,480 tons submerged, almost as big as a light cruiser, with two 5·9-inch guns instead of *U35*'s single 4·1. There were four torpedo tubes for'ard, two aft.

She was a potent weapon, but as she lay off Finisterre after a stormy passage from Germany an armistice was being discussed at German HQ.

Von Arnaud attacked a convoy of ten big merchantmen protected by two auxiliary cruisers and patrol boats. He sank two steamers, but his second victim hit *U139* as she sank, almost taking her to the bottom, and leaving her with a smashed conning tower. She could not dive, but next day intercepted a small freighter carrying port wine and cement. They plugged the hole with the cement, toasted their luck with the port, and continued the patrol, though without periscopes.

Spirits sank when the wireless told them that the Army was falling back in France. The veterans in the *Schwieger*'s crew were bitter at the thought of defeat, after the U-boats' prodigious efforts. They blamed the big ships of the High Seas Fleet, loafing about in harbour when the U-boats were at sea fighting. Off the Azores von Arnaud sank his last victim, an obsolete Portuguese gunboat escorting a large steamer. The little vessel fought a brave delaying action, and the steamer escaped, carrying several US generals.

On 21st October all U-boats at sea were ordered to stop attacks on merchant ships and return to their home ports, to combine with the discredited surface ships of the High Seas Fleet in a last-ditch assault on the British Grand Fleet. The German big ships were to attack ships in the English Channel and the traffic across to France, in the hope that this would also draw the British Fleet down from Scapa Flow across the North Sea, where the U-boats, backed up by a new barrier of mines, would be waiting for them.

Orders to flash up boilers were given on the night of 29th October but the crews of many of the capital ships refused to obey. Destroyers and submarine crews, the seatime sailors, were still loyal. Johannes Spiess, once Torpedo Officer under Weddigen, destroyer of the three *Cressys*, was ordered to take his *U135* into the Jade Roadstead and escort a boatload of marines going to arrest the ringleaders of mutiny aboard the battleships *Thüringen* and *Ostfriesland*.

The two battleships were lying abreast of one another. Spiess steered his U-boat between them so as to be ready to torpedo either of them from his bow or stern tubes. The mutineers surrendered, but the *Thüringen* and *Ostfriesland* were foolishly sent to Kiel, where other dissidents in their crews spread the revolt to the rest of the fleet.

On 6th November Kommodore Michelsen embarked in Spiess's *U135* and led a mixed squadron of loyal submarines, destroyers, minesweepers and other small craft out to sea to find a base from which to continue to protect the German coast. But all likely harbours were under the control of Soldiers' Councils of dissidents, and on 9th November Michelsen reluctantly ordered his ships to return to their home ports.

Two days later the armistice between Germany and the Allies was signed. Among the terms of surrender was the handing over of 176 German submarines. The majority of these reached British or French ports, but eight sank from shame on their way across the North Sea, with no loss to their crews. One of these was Otto Hersing's illustrious veteran *U21*.

On 14th November von Arnaud brought *U139* into Kiel harbour, and Schwieger's men saw the Red flags of revolution flying there.

Von Arnaud, the most successful submariner in history with a score of over 400,000 tons sunk, told his first lieutenant to take over command, walked ashore in his civilian clothes and went home.

Incarcerated in an English prisoner of war camp was young Kapitänleutnant Karl Dönitz. Born at 1892 at Mecklenburg into a Prussian family of landowners and shipowners, he joined the Kaiserliche Marine at an early age and by 1913 was serving in the battlecruiser *Goeben* of the Mediterranean Fleet. He was still in her when she and the cruiser *Breslau* attacked Bône on the declaration of war a year later, then made her epic escape to Turkish waters. Dönitz stayed at Constantinople, where he met the triumphant Hersing after the latter had sunk two British battleships off the Gallipoli beaches. When Turkey entered the war Dönitz took part in raids on Russian ports in the Black Sea.

In 1916 he transferred at his own request to U-boats, and became a watch officer in *U39*. During one of his first actions she was rammed, and with her periscope smashed and her casing torn open, was lucky to make port. He returned to the Mediterranean to take command of *U25*, and in her got into Port Augusta, Sicily, and sank the British repair ship *Cyclops* at the jetty.

Transferred to the *UB68*, on the evening of 3rd October 1918 he made a moonlight attack on a British convoy heading west from Suez.

Dönitz penetrated the destroyer screen unseen and was just shaping up to attack the leading ship of the outside column when the whole convoy changed course and he was nearly run down. Finding himself between the first and second columns he quickly turned the U-boat's head and fired at the ship behind his original target. Dönitz saw the torpedo hit, turned round and made for the vanishing convoy again, remaining on the surface for maximum speed.

He had hoped to be able to make a surface attack in the dark, but it was light when he overtook the convoy, and he had just decided to submerge and attack when *UB68* took things into her own hands. She had always been a rogue boat, with a fault in her longitudinal stability which many modifications by the Germania Yard had failed to eradicate. Suddenly they found themselves diving almost vertically. The battery fluid spilled, all the lights went out, and Dönitz ordered all tanks blown and the rudder put hard over. At three hundred feet, well beyond the official safety depth, then at last the empty tanks began to exert their effect. But the boat was now too light, and the reaction shot her to the surface in an eruption of foam.

All around her were the ships of the convoy. The merchantmen opened up on her with their stern guns, the destroyers came steaming down on her.

Lack of compressed air and soon the shell holes from several hits made it impossible to dive again. Dönitz had no alternative but to abandon ship. All but seven men were rescued and taken to Malta, then transported to Britain, where Dönitz saw the drab insides of several POW camps in England and Scotland. His tough, ascetic, rather prudish nature did not suffer from the actual harshness and frugality of the life, or even the special antagonism of some of his guards to the 'brutal Hun' submarine officers, but for a young active man who, in his own words, 'had been fascinated by that unique characteristic of the submarine service, which requires a submariner to stand on his own feet and sets him a task in the great spaces of the oceans', the restriction of freedom was too hard to bear.

His only hope was to be repatriated as a prisoner who was no longer fit enough for active service. He first tried to fake insanity, with long vacant stares and idiot games with biscuit tins and china dogs, until the Commandant was sufficiently convinced to send him to hospital. To his chagrin he found a ward full of strange muttering German servicemen, some fake, some genuine – it was often hard to tell the difference. He managed, however, to fake a recurrence of malaria, and returned on a stretcher to a defeated Germany, which was expressly forbidden ever to build submarines again, by the conditions of the Versailles Treaty.

CHAPTER EIGHT

Port of Aces

The big tri-motor Junkers with black crosses on wings and fuselage and swastikas on its tail flew low over the farms and fields of Brittany.

Kommodore Karl Dönitz, *Befehlshaber der U-bootë* of the Third Reich, stared down from his cabin window, not really seeing the peaceful scene.

His mind was reviewing ten months of war. Ten months with that constant sense of *déja vu*. Ten months of war with Britain and France. Ten months of U-boats again. The British even had Churchill back as First Lord of the Admiralty, and he himself was back with the boats, though this time he commanded them. Twenty years of waiting, biding time, in destroyers, behind a desk, then command of the *Emden*, a taut ship, for a surface vessel, with the Navy quietly and craftily building submarines again, forbidden fruit after they had almost won the old war for Germany, first the little 250-tonners, the 'canoes', built secretly inside locked sheds in the Deutsche Werke and Germania yards in Kiel, then the 500 and 750-ton boats. Never enough of them, of course. That was an old story too. Fifty-seven boats to start a war with, only twenty-one operational, and he had asked for three hundred.

But Eric Raeder had to have his battlewagons. Well, they hadn't done Germany much good before, and she'd had five times as many then. And what had they done so far? *Graf Spee* scuttled in disgrace, *Deutschland* swinging round the buoy after a furtive dash into the Atlantic, just like von Scheer's useless dinosaurs. This was the day of the aeroplane – and the submarine.

Of course, the Royal Navy had shrunk too. They were desperately short of destroyers, especially after Dunkirk, sloops and patrol boats to protect their convoys – they had remembered their lesson from the last war and started convoy early this time. But they had Asdics and good radar, and they had aircraft carriers, though they had two less of them now, thanks to *U29* and to the one useful job *Scharnhorst* and *Gneisenau* had done so far.

Déja vu again, starting with young Fritz Lemp thinking the *Athenia* was an auxiliary because she was steaming without lights – the *Lusitania* all over again. And Otto Schuhart bagging the *Courageous* . . . Günther Prien knocking off the *Royal Oak* in Scapa Flow . . . What were they if they weren't the *Cressys* of the new war? Of course, he'd done better than Heinrich von Hennig, and poor Hans Emsmann who had lost his boat and his life in the Flow in 1918, for nothing.

First, six months of what the Americans called the 'phoney', the British the 'twilight' war. Twilight for them, perhaps, this time, though there had been occasions when he'd wondered with foreboding whether history was repeating itself . . . Then suddenly the pattern was broken. Holland, Denmark, Norway were all theirs, the Schlieffen Plan had worked the second time round, and France had fallen like a rotten fruit: this time a Contemptible Little Army was cowering on the beachs of Dunkirk, and the whole western coast of Europe was German, from the North Cape of Norway to the Spanish border. Now Göring's Luftwaffe was poised on all the coastal airfields, and all the ports belonged to the Kriegsmarine. The Norwegian harbours of Narvik, Bergen, Trondheim and Christiansand, and the French ports of Brest, Bordeaux, La Rochelle and St Nazaire, with Lorient as the headquarters port, were all to be U-boat bases.

The heavy plane rolled to a stop on the dusty airfield. Men in field-grey pushed the high steps up to the fuselage door, the door opened and the short wiry figure of the Flag Officer, Submarines, in his dark blue uniform with its sky-blue lapels stepped briskly down, returned the salute of a Luftwaffe major at the foot of the steps, and walked the few yards to the waiting Mercedes staff car, its top down in the hot July weather.

The little town of Auray looked almost untouched by war as they sped through, but alongside the roadsides were abandoned British and French army trucks, civilian cars, motor cycles.

These signs of flight and retreat continued all the way into Lorient. Dönitz's car drove into the square in front of the Préfecture, the new Flotilla Headquarters, where Kriegsmarine officers and ratings were coming and going.

Lemp's *U30*, which had been the first German submarine to tie up in the captured French Biscay port, was in from patrol. It was very strange, that fleshy young pirate told Dönitz, the *Athenia* affair forgiven and forgotten, to finish a patrol off the Scillies and, instead of the long, weary haul up round the Irish coast and through one of the gaps in the Scottish islands back across the North Sea to Wilhelmshaven, to head *south*, past Ushant and the unfamiliar landmarks of Pointe de St Mathieu, Pointe de Raz, Pointe de Penmarch, and east to leave the Ile de Groix to starboard and up into the twisting Scorff Estuary, past Port Louis into the new berth. Doubtless it would all soon become as familiar as the Schillig Roadstead.

When *U30* had arrived on 7th July, HQ had already been set up in the Préfecture, and the Hotel Beau Séjour taken over for officers' quarters. It was a comfortable billet, and there were a hundred thousand bottles of good vintage champagne in the cellars. Crewmen were rather less cosy in the nearby School of Military Music. Rest camps had been established at Quiberon and Carnac, a few miles further south.

As the boats finished their patrols in July they put into Lorient for repairs by workmen flown in from the Germania yards at Kiel.

In mid-July Kapitänleutnant Otto Kretschmer brought his *U99* into Lorient. An exhausted, unshaven crew trooped ashore and headed for the comforts of a soft bed, a bottle and female company, in that order. It had been a tough patrol, with some very close depth-charging, and their need for rest and relaxation was desperate.

Other boats, battle-hardened after a cruel winter in which half a dozen U-boats had sunk well over a hundred Allied ships, more than half a million tons, made their first landfalls off the Breton ports. There was Vaddi Schultz, Knight's Cross, first man in this war to sink 100,000 tons, and young commanders who, having done exceptionally well in the canoes, had been promoted to bigger boats. The tall, dark, handsome, swashbuckling Joachim Schepke with the little *U19* and only five torpedoes had in one patrol survived damage from heavy depth-charging to sink four ships, a bag of over 20,000 tons, and had been rewarded with the command of *U100*.

Prien, whose earlier adventures in Scapa Flow had caused the British Home Fleet to be packed off to the west coast of Scotland, had had a spectacularly successful patrol in June, sinking ten ships, a total of 66,588 tons.

U47 climaxed this record score by torpedoing the liner *Arandora Star* without warning three hundred miles north-west of Ireland on 2nd July.

A Royal Air Force Sunderland flying boat on patrol answered the liner's SOS signal, and found her at 11 a.m. in a sinking condition.

The aircrew counted thirteen lifeboats bobbing about in the cold choppy sea, with other survivors clinging to scattered odds and ends of wreckage.

The big four-engined flying boat flew low, and the crew dropped all the food they had on board, including their own emergency rations and all their cigarettes and tobacco, which they put in waterproof bags and lashed to their own life jackets to act as buoys. Having made sure that most of the packages had been picked up, they flew off to find a ship. They quickly picked up a destroyer and guided her to the survivors, who were rescued about three hours after the Sunderland's first sighting them.

While the destroyer's boats pulled round picking up the shocked, exhausted and wounded people, the Sunderland remained overhead, prowling vigilantly from boat to boat, from individual to individual, dropping smoke flares and firing Very lights to make sure that the destroyer found every last one.

After about an hour of this work, the destroyer flashed to the flying boat.

'They are mostly German internees.'

On board the *Arandora Star* had been 1,500 Germans and Italians bound for internment in Canada.

The Sunderlands hit back at the U-boats. Five attacks in five days were recorded in June, including one by an Australian crew, whose skipper afterwards said, 'He was cruising on the surface, and he saw us at about the same time that we saw him. He dived but we were over him while he was still at periscope depth, and we could see him silhouetted under the surface as we dropped six bombs close to the port bow. A mass of bubbles and oil came up.' Some days later a signal went out from U-boat Headquarters at Lorient, '*U53*, report your position . . *U53*, report your position . .' No reply was received, and an asterisk was put against her number in the list at Sengwarden, followed after a further interval by a second asterisk. That was final. After that the *Befehlshaber der U-bootë* himself always wrote personally to all bereaved wives and parents.

On 19th July Britain publically rejected Hitler's last peace overture. The Führer activated Operation Sealion, the invasion of Britain. 'We shall fight on the beaches', promised Winston Churchill, now Prime Minister. The Luftwaffe on the French Channel coast stepped up their pounding of Allied shipping in the narrow waterway.

After a final drunken run ashore, *U99*'s crew rejoined her on 24th

July. Otto Kretschmer, twenty-eight, the son of a schoolmaster, was a quiet, reticent, almost taciturn officer who did not indulge in the flamboyant and flashy behaviour of some of the other young U-boat captains. His crew called him 'Otto the Silent', and he was a strict disciplinarian, though by no means unpopular with them. He was fair, and, above all, a fine seaman and submariner, who always knew exactly what he was doing. His continued success at sea rubbed off on his men, and he was wise enough to let them have their fling ashore. They boasted about him in the estaminets of Biscay.

Before they shoved off, Otto the Silent had unbent enough to allow the boat to acquire a logo. The ship's artist had painted on the conning tower two golden horseshoes, ends down.

Two weeks later the golden horseshoes came in sight again off Port Louis. Flying from *U99*'s periscope standard were seven victory pennants, with the horseshoe device emblazoned on them as well, signifying a record total of 65,137 tons sunk in one patrol, which was cut short only when the torpedoes were used up.

Grossadmiral Raeder himself was on an inspection of the Lorient base when *U99* came in. He looked with astonishment at the unshaven crew's assortment of abandoned British Army and RAF uniforms, but said nothing in view of their outstanding achievement. With the band playing and ranks of German soldiers cheering, he conferred on Kretschmer the Knight's Cross. Before long Lorient was being called. 'The Port of Aces'.

Britain was now heavily besieged. Many merchant ships had been fitted with some sort of gun or anti-aircraft device, but there were no escorts. On 31st July Winston Churchill cabled to Mr Roosevelt: 'It has now become most urgent for you to let us have the destroyers, motor boats and flying boats for which we have asked . . . '

On 10th August came *Adler Tag* (Eagle Day), when Göring unleashed his Stukas and Me109s in great strength against the radar stations and airfields of southern England, to destroy Fighter Command of the Royal Air Force as a preliminary to invasion.

On 17th August Germany declared the whole of the area round the British Isles north to 60°and west to 20° a war zone, in which ships could be sunk without warning. It was officially unrestricted submarine warfare once again.

On 25th August Schepke in *U100* approached Convoy HX65 on the surface in darkness, and sank three ships. Three days later it was outward-bound OA204's turn. Even if spotted in the darkness Schepke had little fear of the convoy's escort of one corvette and one armed trawler for twenty-one ships. Just before midnight *U100* fired a 'fan' of torpedoes when the convoy was 175 miles off Bloody Foreland, Northern Ireland. The first missile hit the British SS *Hartsmere* on the starboard side under the bridge. One minute later another tinfish struck the Commodore's ship SS *Dalblair* amidships, and she sank in ten minutes.

Dönitz still had only fifty-seven boats on strength, exactly the same number as at the outbreak of war, the twenty-eight new boats having exactly equalled the number of losses, but Italy had joined the Axis in June, and early in August twenty-seven Italian submarines arrived at Lorient for duty in the Atlantic. In August they sank only four ships, and Dönitz sent them to the Azores, where there was no opposition. Here they sank three ships in the month of September.

On 31st August Roosevelt cabled to Churchill, 'It is my belief that it may be possible to furnish to the British Government as immediate assistance at least fifty destroyers'. Most British destroyers were held in the Channel by the invasion threat, and there were only fifty escort vessels to protect the convoys.

In the circumstances Dönitz thought that it was time to begin experimenting with his *Rudeltaktik*, or Pack Attacks, which his boats had practised in the months before the declaration of war. In this form of attack, which had been tried half-heartedly in the Great War, a 'pack' of from six to ten U-boats formed a patrol line across the likely path of a convoy. Whichever boat first sighted the enemy did not attack alone, but continued to shadow and report the position and course of the convoy to HQ at Lorient, which homed the rest of the group on to the ships. When they had all closed the convoy, they waited for darkness, then used their diesels, which gave them a speed superior to that of the convoys, to attack in unison on the surface.

On 1st September he moved his headquarters from the barracks at Wilhelmshaven to a spacious house on the Boulevard Suchet in Paris as part of a centralization plan for Operation Sealion. But the Luftwaffe's attempt to destroy RAF Fighter Command had failed. On the night of 7th/8th September Göring's bombers switched their attention to London. The Battle of Britain was almost over. The full fury of the Battle of the Atlantic was about to begin.

On the night the Luftwaffe changed its targets, a quartet of U-boats, proceeding on information provided by the German Radio Intercept service, which was becoming quite successful at the decrypting of Allied signals, ambushed the fifty-three ships of the slow homeward-bound Convoy SC2. Kuhnke in *U28* was the first to sight the convoy. In very heavy weather he held on and guided Prien and Kretschmer to the attack. A Sunderland kept them down for a time, but a co-ordinated attack sank five ships.

Ten days later Heinrich Bleichrodt's *U48*, the boat with the arched black cat emblem, ranging six hundred miles out into the Atlantic, sighted the 11,000-ton passenger liner *City of Benares*, which was carrying a hundred children on a special evacuation scheme to North America.

It was 10 p.m., and most of the children were asleep when he torpedoed her. She sent out urgent SOS messages, and ships and aircraft raced to her assistance. For some days many of the passengers and crew remained unaccounted for, and there were misgivings about their safety, as the seas were very rough.

Then, on 25th September, a Sunderland of Number 10 (RAAF) Squadron sighted a ship's lifeboat with a number of people in it about two hundred miles closer to Ireland than the position of the *City of Benares* sinking. There were forty-six people in the boat, including six children. Thirteen children were rescued from the *City of Benares* in all. Seventy-seven died, victims of the Arched Cat.

Meanwhile Günther Prien was still at sea, with one torpedo left after the savaging of SC2. The crew of *U47* were looking forward to the soft beds of Lorient, and were not pleased when they were diverted far out into the Atlantic on a weather patrol.

The Bull of Scapa himself was hunched on the bridge, dodging the spray and eating salt, when he sighted smoke, then masts, on the horizon ahead. Soon he was counting the forty-one ships of the fast inward-bound HX72.

He signalled the position and speed of the convoy to HQ, then clung to it while the air hummed with homing signals. Next morning Kretschmer, Schepke and Bleichrodt joined him, and that evening the four aces closed in for the kill on the surface, manoeuvering to put the bulky silhouettes of the merchantmen against the bright moonlight. The sea was calm, though occasional heavy rain showers spattered the bridge and necessitated oilskins.

HX72 had a minimal, inexperienced escort of four ships, led by the sloop *Lowestoft*. About 10.20 p.m. the red light signalling that a ship had been torpedoed soared into the night from the centre of the convoy. The Senior Officer Escort took *Lowestoft* first across to the port flank, where a bright shower of starshell revealed nothing but empty ocean, then round towards the rear of the convoy, in company with the corvette *Heartsease*. Then a ship was torpedoed front-centre of the convoy, another in the rear of the port column.

The convoy began to break up, and the sea wolves fell upon the ships. More red flares rose up like cries of despair. By signal and loud-hailer SOE tried to round up his charges but they were too scattered.

With the U-boats attacking on the surface, all the escorts could do was fire starshell to try to pick them out, but the submarines were faster than they were. Kretschmer and Co were grateful for the extra illumination, which made targets easier to see. The few escorts were slow and easy to dodge. Only once in seven hours of attrition did an escort's asdic get a ping and start a DC attack. In that time eleven ships were sunk and 100,000 tons of supplies from the USA lost.

In October Hitler postponed Operation Sealion, and the Luftwaffe began bombing British ports and factories at night, with U-boats and surface raiders stepping up their campaign of attrition at sea. Dönitz moved his headquarters from Paris to a villa requisitioned from a sardine merchant at Kernavel, between Lorient and Lamorplage. From his operations room here he had a view of Port Louis and the fort at the entrance to the harbour, and could see any of his boats returning from patrol in plenty of time to drive down and meet her at the quayside, which he invariably did.

Just before midnight on 16th October Heinrich Bleichrodt's *U48* sighted the big slow convoy SC7 west of Scotland, bringing American supplies to Britain, escorted by three old sloops. The Arched Cat sank the new 10,000-ton French motor tanker *Languedoc* and the freighter *Scoresby*, and alerted headquarters, which ordered five other submarines, patrolling to the north and east of Rockall, to head for the convoy, Fritz Frauenheim in *U101*, Karl-Heinz Moehle in *U123*, Engelbert Endrass in *U46*, Joachim Schepke in *U100*, and Otto Kretschemer in the well-tried *U99*. They all converged on the last position received from *U48*, awaiting further reports from her as she shadowed the convoy.

But Bleichrodt was diving fast, Sunderland's bombs exploding close aboard him, lights out, dials shattered, water spraying in through the hatches and periscope gaskets. SOE's sloop *Scarborough* left the convoy on a fruitless hunt for *U48*, and never rejoined the convoy, though she did prevent Bleichrodt from continuing his shadowing reports.

When *U48*'s signals to Lorient ceased, the five boats sent to close the convoy were left with too big an area to search. They were ordered to reverse course, head east and form a north-south barrier east of Rockall, well ahead of the convoy's last known position, as reported by *U48*, with the hope of sighting it on the morning of the 18th. Meanwhile, Liebe's *U38* and Kuhnke's *U28* were alerted in case the convoy passed through their areas. At dusk Heinrich Liebe sighted SC7, radioed its position and course to Lorient, torpedoed the freighter *Carsbreck*, then resumed his original patrol.

At 8.15 p.m. Engelbert Endrass' *U46* fired a salvo of three torpedoes at the unguarded port bow of the convoy, sinking the Swedish *Convellaria*. Two more sloops had joined the convoy, making four, still ludicrously insufficient for a convoy of thirty-four ships, and the wolf pack attacked. Most of the boats followed doctrine and fired 'fans' of torpedoes from outside the convoy. Kretschmer had his own philosophy of 'one ship, one torpedo', and made individual attacks wherever a target presented itself.

Twenty ships from SC7's original thirty-four were sunk, seven by Otto Kretschmer, four by Moehle, the remainder by Endrass, Frauenheim and Schepke between them, and valuable cargoes of steel, ammunition, timber, oil, petrol and aircraft were destroyed. Their bloody and brilliant action came to be known as 'The Night Of The Long Knives'.

The following night of 19th/20th October was almost as grim for the forty-nine ships of Convoy HX79, steaming close on the heels of the shattered SC7. This important convoy was making a 'fast' eight to nine knots, and picked up a stronger escort at the rim of the Western Approaches, comprising two destroyers, four corvettes, a minesweeper and three anti-submarine trawlers, which joined the ocean escort of two armed merchant cruisers. There was even the Royal Dutch submarine *0-14* sailing in the middle of the convoy, and the 5,000-ton *Loch Lomond* was to act as rescue ship for survivors of torpedoed vessels. But as with SC7, there was no plan to counter surface pack attack at night.

Günther Prien in *U47* had missed SC7 but it was he who first sighted HX79 in the morning, and radioed its position and estimated speed immediately to HQ.

Kretschmer, Frauenheim and Moehle had all used up their torpedoes and were on their way back to base, but Endrass and Schepke still had some shots in the locker, and Liebe's *U38* and Bleichrodt's *U48*, with only one serviceable torpedo left, were also roped in by Lorient to supplement the attack on HX79. Bleichrodt had had one more brush with *Scarborough*, when he had sighted the

U-boat surfaced some way off, but the Arched Cat had run too fast for the old sloop.

That evening the four U-boats gathered round HX79 and attacked as night came down. The result was like a re-run of SC7's ordeal, a second Night Of The Long Knives. The first to go was the 6,000-ton motor tanker *Sitala*, and twelve more ships followed her, the last to be sunk being the rescue ship *Loch Lomond*, which had in any case been unable to cope alone with the hundreds of survivors. Endrass and Schepke each sank three ships. Even then the slaughter of the merchantmen was not over. The same night a third convoy, HX79A, was ambushed by Dönitz's boats, and seven ships lost.

Problems of sighting the convoys would have been improved for the U-boats with good air reconnaissance. Since late June of the year the anti-shipping unit Kampfgeschwader 40 had been operating from the airfield at Bordeaux-Mérignac, using Focke-Wulf Fw200 Condors. These big four-engined machines had been adapted early in the war from a civil airliner designed specifically to satisfy the request of the Kriegsmarine for a long-range aircraft capable of making armed reconnaissance sorties in co-operation with its submarines well out in the Atlantic, the specially designed Heinkel He177 having developed serious teething troubles. As a derivative of a civil aircraft, the Condor was vulnerable to anti-aircraft gunfire and the extra stresses of combat flying, but it had a radius of action of nearly a thousand miles, carrying a normal load of four or five 250 kg (551 lb) bombs, and had proved very effective as a bomber against single, unescorted, poorly armed merchantmen. With only some seven or eight aircraft on strength at any given time, and only three or four of them operational, the bombing effort had given them little or no time to recce for the U-boats.

Co-operation between the Navy and the Luftwaffe was notoriously bad in any case, made worse by a personal antagonism between Reichsmarshall Hermann Göring and Dönitz, who were chalk and cheese. Time and again the *Befehlshaber der U-bootë* complained that his boats were not getting the assistance needed from the airmen. Then KG40 would grudgingly release one machine to work with the submarines for a time, though so far there had been no instance of a Condor directing U-boats to a convoy, and if any aircraft sighted a straggler it usually had first go with its bombs from masthead height.

This was what happened in the first phase of an attack on 26th October. At 4 a.m. that Saturday the Condor of Oberleutnant

Bernhard Jope lifted off the runway at Bordeaux-Mérignac for 'armed reconnaissance over the Atlantic to the north and north-west of Ireland', and headed out over the sea in the pitch dark. Five hours later, a few minutes after nine o'clock, they were butting through heavy cloud and rain about seventy miles north-west of Donegal Bay when three huge funnels became visible through the greyness of the drizzle below.

The 42,500-ton Canadian Pacific liner *Empress of Britain* was on the last leg of her voyage from the Middle East, carrying Servicemen and their families. This prestigious and beautiful ship had a top speed of twenty-four knots and was making the passage unescorted.

The big Condor dived, flew over the ship from stern to bow at five hundred feet and dropped a bomb which hit the *Empress* abreast the centre funnel. The upperworks amidships began to blaze fiercely and give off clouds of choking, blinding black smoke, which obscured the vision of the ship's after 3-inch AA gunner, as she was not steaming into the wind. The Condor made two more runs, the after steering position was hit, and the 3-inch wrecked. A dense pall of smoke rose up, fire consumed the exotic public rooms and glowed out of portholes and hatches, and the great ship listed to starboard. Jope was sure she was doomed and set course for Bordeaux.

By heroic efforts most of the people aboard this floating holocaust were taken off by the few boats that it had been possible to get into the water. These were transferred to destroyers and armed trawlers. The *Empress*, still smoking fore and aft, was practically gutted but was still afloat on an even keel, and the tugs *Seaman* and *Thames* were ordered to salvage her.

Jope's attack report to KG40 at Bordeaux had been passed on as routine to U-boat HQ at Lorient. German Intelligence and further air reconnaissance established the identity and approximate position of the Condor's victim. In the afternoon the 500-ton *U32*, commanded by Oberleutnant-zür-See Hans Jenisch, an experienced submarine skipper with over 100,000 tons of Allied shipping in the bag, which had left Lorient for her operational area north-west of Ireland the day before, picked up a radio signal from Lorient informing all U-boats in the area that the English liner *Empress of Britain*, now being used as a troop transport, had received bomb damage from a German aircraft about three hundred miles west of Malin Head.

The reported position, which was inexact, lay off *U32*'s course of her patrol area, and *U31* was likely to be closer to the ship, so Jenisch did not act on the report. But when a second message came through next day to say that the *Empress* lay burning and crippled in the same position Jenisch altered course towards her, steering sixty miles to the east of the reported position, in case she managed to make any way towards home, if she had not sunk in the meantime.

Towards noon next day, with visibility good, they sighted the trucks of the *Empress's* masts, dead ahead. Other masts could be seen, including those of escorting destroyers, as well as aircraft overhead. *Empress* was under tow, escorted by the destroyers *Broke* and *Sardonyx*.

The presence of a Sunderland forced *U32* to dive and kept her below all day. With dusk approaching, Jenisch surfaced but could no longer make out the target, and submerged again to use the hydrophones. At a distance of twenty miles they picked up propeller noises, surfaced and headed towards the position at a good speed. Their target was steaming very slowly, about four knots, and towards midnight the *Empress* was sighted, with her destroyer escorts off her port and starboard bows.

Jenisch followed the ships on the surface for about two hours, assessing their course, speed and escort pattern. The destroyers were maintaining a zigzag, and there came the perfect moment, when they moved outwards from the *Empress* and opened up a gap into which the U-boat could slide for a point-blank attack.

Jenisch seized the opportunity, manoeuvered the boat into position to port of *Empress*, and dived to fire two torpedoes from a range of between 550 and 600 yards, one aimed at the ship's foremast, one at her mainmast, aft, then turned away.

The first torpedo detonated prematurely. Jenisch turned *U32* towards the target again and aimed a third torpedo at her middle funnel. The liner was huge in the periscope's eye. The premature explosion of the first torpedo had obviously alerted the skeleton crew aboard *Empress*, and lights moved about her scarred and blackened decks, sirens hooted. Her foc'sle was still on fire.

One after the other the two torpedoes found their marks, rocking *U32* with the explosions. One hit amidships, and a boiler exploded. A white mushroom cloud rose high over the ship, which listed rapidly to port. *U32* was so close that in the periscope the huge shape seemed to be falling on top of them.

The tugs cast off their lines, and the destroyers probed for the U-boat with their searchlights in the likely firing position off the *Empress's* port bow. *U32* withdrew swiftly, trimmed low, down the wake of the liner, and Jenisch watched the destroyers circling the toppling giant, firing at shadows on the sea. A flying boat swept very low over the submarine's conning tower without seeing them.

Empress's list increased, and after ten minutes she sank. As a statistic she was just one of sixty-three ships sunk by U-boats in October, totalling 352,407 tons, the highest monthly loss of the war thus far. Her sinking was part of 'The Happy Time', as the U-boat service called the four months after the occupation of the Biscay ports.

Jenisch set off westwards and reported the sinking to Dönitz. Two days later *U32* was sunk by the destroyer HMS *Harvester*, after an attack on a convoy straggler. Jenisch and most of his crew were taken prisoner.

Otto Kretschmer had now emerged as the top-scoring U-boat 'ace', and on 3rd November he capped his achievements. Before the beginning of war the Admiralty had selected fifty passenger or cargo liners ranging from some 6,000 to 20,000 tons, with a top speed of at least fifteen knots, and armed them with a scratch collection of old 6-inch and 4-inch guns, some of these dating from 1898, to serve in lieu of the regular cruisers which the Royal Navy lacked. These unprotected, poorly armed auxiliaries, with their high silhouettes, were very vulnerable to attack, and five of them had been sunk by U-boats since June of the year, with one seriously damaged. Many of them had served usefully on the Northern Patrol blockading Germany, and after the entry of Italy into the war in June some had been switched to the Western Patrol watching for Italian ships off the Straits of Gibraltar.

On 3rd November the 18,724-ton Cunard White Star liner *Laurentic* was returning home after Western Patrol duty. She had been laid up for two years before being coverted to an AMC, had already suffered heavy weather damage with the Northern Patrol, and had caught the blockade runner *Antiochia*. In company with her was the China Mutual Steamship Company's *Patroclus*, a coal burner.

At 9.45 p.m. Captain Vivian was in the *Laurentic*'s chart room when she received a distress signal from the SS *Casanare*, which had been torpedoed about thirty miles away.

The *Laurentic* was a big ship, with a high hull and very tall funnels, and on a clear night such as this, was visible at a distance of up to eight miles. She too was a coal burner, and it had never been possible to keep her smoke down to really safe proportions. The ship was zigzagging at fifteen knots, and the navigator had just shown the captain their position on the chart when they were hit by a torpedo on the starboard side. They had been a perfect target for *U99*.

Captain Vivian pressed the alarm gong buttons for Action Stations. The two guns' crews on watch manned one forward 5·5-inch and one 4-inch. The ship had been hit in the engine room, and was listing to starboard.

The sea was calm, and the captain gave the order to turn out all boats, with everyone except the two guns' crews to go to their Abandon Ship stations. He then saw the *Patroclus* coming up and warned her that he had been torpedoed by firing a series of red Very lights.

As these fireworks were blooming in the night sky, the men on the *Laurentic*'s bridge sighted a submarine surfacing on their starboard bow. Captain Vivian shouted to the forward guns to open fire, but someone had apparently given the order for the crews to fall out. He sent Midshipman Nicholson to get the after 4-inch into action but the training was jammed. He himself was hurrying to the starboard forward gun when they were hit by a second torpedo. The shock freed the after 4-inch, and both guns fired starshell, then high explosive, and the U-boat dived. Captain Vivian went round the ship and made sure that everyone had got away, except for three men who refused to leave; then he climbed down into the Carley float and cast off. Shortly afterwards there was a third explosion, and the *Laurentic* sank by the stern.

The *Patroclus* had come up with the stricken *Laurentic* and circled round the position for some time until Captain Wynter judged it sufficiently dark for them to slow down and pick up survivors. As she came in, the *Patroclus* dropped two depth-charges set to 150 feet to scare away the U-boat or U-boats in the vicinity.

They saw boats in the water, and hailed the nearest one, *Laurentic*'s Number 6. This boat was just coming alongside the ship at her forward welldeck when a torpedo struck the *Patroclus* right underneath the boat, blowing it to pieces. Few men were saved from the boat, and some *Patroclus* men were killed instantaneously by the explosion. Some were blown overboard, and the welldeck became a shambles.

The rest of the ship's company were assembling by their boats when another torpedo hit the *Patroclus* in Number 4 Hold. Captain Wynter ordered 'Abandon ship' and the magazines flooded. The ship now had a big list on her, and launching boats in the dark was difficult, but most of them got clear. Then two more torpedoes struck the ship simultaneously, blowing S3 gun completely overboard. Captain Wynter put the telegraphs to Finish With Engines, and ordered all hands into the water.

All boats and Carley floats had left the ship, but there were still several officers and ratings left aboard when the U-boat surfaced and started shelling the ship.

As two shells hit the *Patroclus*, they ran to man the nearest gun, the starboard 3-inch AA, and opened fire, with Chief Petty Officer Creasey as gunlayer, Able Seaman Ellis as trainer, the First Lieutenant, Commander Ralph Martin, as loader, and Lieutenant Commander Hoggan as ammunition supply. They fired four rounds, Creasey and Ellis reported them dead on target and thought they had hit with the fourth round. At any rate, the submarine disappeared, her shells having started a fire on the after welldeck over the magazine. Ten minutes later, about quarter past one in the morning, another torpedo hit the ship in Number 3 Hold. Still the *Patroclus* did not sink.

About 4 a.m. the helpless ship shook with the impact of a sixth torpedo, underneath the bridge. The bridge collapsed completely. Martin said, 'The silly bugger has put four tinfish in the same place,' and they all laughed. This hit practically broke the back of the ship forward of the bridge. Five minutes later came yet another hit amidships in the engine room. The ship began to list steeply and they realised the end had come at last. Martin said, 'Come on, over the side.' They had already attached lifelines inboard from the davit heads, and now slid down these into the water.

They swam away, felt the suction of the sinking ship pulling at them, and heard 'a terrific noise as all the girders twisted and bent . . .' The ship rolled over and her bows rose in the air, but she did not sink for another two hours. Several men died in the water, then the destroyer *Achates* came up and picked up Martin and the survivors, breaking off once in the middle of the rescue on an abortive asdic hunt.

Lieutenant-Commander David Wanklyn, RN, of *Upholder* (left) and his First Lieutenant, Lieutenant Crawford.

HMS *Ursula*, one of the first U-class submarines to operate from Malta, commanded by Lieutenant-Commander A. J. Mackenzie.

HMS *United* operated from Malta in 1942.

A kill by *U960* (Oberleutn[a]
Heinrich) on 7th February
1944.

CHAPTER NINE

Trumping the Ace

Jope's Condor attack and *U32*'s successful follow-up had set an example of co-operation between airmen and submariners.

On 30th December 1940 Dönitz said at a combined forces staff meeting, 'Just let me have a minimum of twenty Fw200s solely for reconnaissance purposes, and the U-boat successes will shoot up.'

The Allies were picking up U-boat signals by the use of HF/DF (High Frequency Direction Finding) and re-routing convoys out of the way of his boats. If he ordered a reduction in radio traffic he would lose touch with the boats and be unable to exercise the degree of tactical control he wanted. All he could do then was direct them to areas of likely convoy movement.

It was not good enough. But as for twenty Condors . . . He would be lucky if Peterson allocated the usual miserly one. When he complained, Raeder, although sympathising with him, merely referred him to Jodl. So on 2nd January Dönitz went to see General Jodl, head of the operations division of the Supreme Command. Rather more realistically he stressed the need for twelve maritime recce machines in the air at any one time, omitting to mention the back-up strength this would involve.

Jodl too was sympathetic, and on the 4th he was telling Hitler that to enable their naval command centres to prosecute the war in the Atlantic, systematic reconnaissance was essential.

Two days later, to Dönitz's amazement, Hitler decreed: 'I Gruppe/KG40 will be under the command of the Commander-in-Chief of the Navy.'

Dönitz's reaction was that it was fortunate the 'Fat One' was on leave, as he would never have stood for it.

He found that the KG40 'Gruppe' was much weaker than he had thought, with just twelve Condors on strength, and an average serviceability of no more than six aircraft, operating one or two sorties a day over the Biscay area and to the north and west-nor'-west of Ireland. On 8th January the Blue Funnel line's motorship *Clytoneus*, on her way alone from Belawan to Liverpool with tea, rubber, spices and palm oil, hit an attacking Condor with her 12-pounder gun, and the aircraft crashed in France. Two days after this the armoured ocean-going rescue tug *Seaman* shot down another Fw200 in the sea. Dönitz developed some sympathy for Oberst Edgar Peterson, CO of KG40 and originator of the military Condor, who, like him, had been complaining about having to fight on a shoestring.

A convoy when sighted by a Condor had first to be reported, then shadowed, whether attacked or not. There was no direct communication between Condors and U-boats. The sighting Condor radioed KG40 at Bordeaux, who informed the Luftwaffe Liaison Officer at U-boat HQ in Lorient, who in turn passed on the information to the Navy. The U-boats nearest to the targets were then contacted. On 27th January a Condor located OG51, a Gibraltar-bound convoy, attacked and badly damaged the *Baron Renfrew*, carrying machinery castings to Huelva in southern Spain, and destroying the loaded collier *Pandion*. As a result of the Fw's sighting report a U-boat was directed to the convoy and sank a straggler, the small motorship *Pizarro*.

At sea at this time was the Q-ship *Crispin*. The Admiralty had experimented with the 'decoy' ship, many of which had had individual successes in the Great War, but the early general arming of merchant ships and adoption by U-boats of sinking without warning made them of limited use and few were commissioned. The *Crispin* was with her sixth convoy, OB280: she left on the night of 3rd February in company with three escort vessels to join a homeward-bound convoy, and was eight miles from OB280 when she was torpedoed by a submarine and sunk.

On 3rd February Hermann Göring returned from his hunting trip and was furious to find that his Condor unit had been handed over to the Kriegsmarine. He invited Dönitz to a meeting to discuss the transfer.

Early after dawn on 4th February a Condor picked up a homeward-bound convoy and reported its position. The cumbersome chain of communication creaked into action, and *U107* was vectored to the convoy. Meanwhile, the Condor bombed a straggler, the 4,443-ton Greek freighter *Calafatis*, carrying steel and nitroglycerine from Philadelphia, and broke her in two. Next morning another Greek straggler, the smaller *Ionnis M. Embiricos*, was sunk, but a Condor which attacked the SS *Major C.* was hit by gunfire and crashed at Schull, County Cork, in Ireland, killing five of her six-man crew. *U107* found the convoy and sank the 5,358-ton *Empire Engineer* and the 3,388-ton straggler *Maplecourt*.

This victory for co-operation cut no ice with the *Reichsmarshall*, who received the *Befehlshaber der U-bootë* in his private train at Pontoise, near Paris, on 7th February. He demanded his aircraft back. The lean, ascetic Dönitz disliked everything about the *Reichsmarshall*. Frigidly he replied that he had the *Führer's* directive.

On the 8th, the day after this meeting, the Kriegsmarine was able to return the favour done them four days earlier. That evening Nicolei Clausein in *U37* intercepted the convoy HG53 half-way between Portugal and the Azores, en route from Gibraltar to Britain. He sent a sighting report, which was forwarded to KG 40 via Flag Officer, Submarines, at Lorient, where Dönitz had earlier that day returned from Pontoise. In the early hours of the 9th *U37* attacked, sinking the freighters *Courland* and *Estrellano*. Clausein then continued shadowing the convoy, reporting its position, course and speed.

At 6 a.m. five Condors took off from Bordeaux-Mérignac, led by Hauptmann Fritz Fliegel, CO of 2 Staffel. They found the convoy at noon some four hundred miles south-west of Lisbon. Sixteen ships steamed along in ragged lines, protected by nine escort vessels. Clausein watched the Luftwaffe in action with admiration, and when they had done their worst they torpedoed the 1,473-ton *Brandenburg*. Between them aircraft and U-boat had sunk eight ships, half the convoy's strength. And Clausein was not yet finished. On 11th February he sighted the slow, unescorted Sierra Leone Convoy SLS64 of nineteen ships east of the Azores. He sank several of the ships himself, and his report enabled the heavy cruiser *Admiral Hipper*, which had broken out from Germany and was operating from Brest, to steam flat-out through the night, reach the convoy at daybreak and sink seven ships, before bad visibility hid the rest.

The band played on the quayside, and half the Germans in Lorient were there to wave *auf wiedersehn* when Prien, in a new leather jacket, departed in *U47* on the morning of 20th February.

Two days later *U99* completed storing and ammunitioning ship, and Kretschmer followed him out, with a band playing the Kretschmer March, specially composed by the Army bandmaster in his honour. By this time Prien had sunk 245,000 tons of Allied shipping, Schepke 230,000 tons, and Kretschmer 282,000 tons. The three leading aces had an agreement that whichever one reached his 300,000 tons first should be wined and dined by the other two. This patrol could well be a fast gallop down the finishing straight, for Schepke in *U100* had come out from Wilhelmshaven. Kretschmer's crew were counting on their golden horseshoes to get *U99* first past the post.

Early in the morning of 26th February Prien sighted outward-bound convoy OB290, and at 1.30 a.m. began his attack, sinking three ships totalling 15,600 tons and damaging three more. His sighting report brought six Condors to the scene, which attacked singly and in waves throughout the day. They sank nine ships from the convoy and damaged two more.

Late on 6th March Prien sighted smoke, and at 4.24 a.m. on the seventh signalled the position, course and speed of a convoy. This was OB293, outward-bound from Liverpool. *U88*, *U70* and *UA* (built in Germany as the Turkish *Batiray*) headed for this one, but were put under and heavily depth-charged by escorts, the destroyers *Verity* and *Wolverine* and the corvettes *Arbutus* and *Camelia*. After two hours of this pounding Kretschmer surfaced. He was unscathed, but at 6.50 Matz reported damage to his conning tower, and Eckermann signalled that he was too badly damaged to continue the attack. Nothing was heard from Prien's *U47*.

Just before dusk, two hundred miles south of Iceland, Prien had sighted an inward-bound convoy. Heavy rain hid it from the enemy, then the rain squall cleared to reveal *U47* to Commander Jim Rowland in the destroyer *Wolverine*. Prien threw the helm hard over and tried to get away on the surface. But even an old V-and-W could outrun a submarine, and Rowland went to full speed in the chase.

Seeing the destroyer gaining on them, Prien dived. The *Wolverine* reached the swirl and dropped DCs. Columns of white water shot up, and the hydrophone operator in the *Wolverine* began to pick up a new rumbling sound, which Rowland thought could be the sub's props running out of alignment, straining shafts and engine mountings. The *Verity* joined him in alternate DC attacks. As night came down Prien surfaced again and went to full diesel speed, or as full as his damage allowed, relying on the darkness to hide him. But the noise his wobbling props made came over loud and clear on the destroyer's hydrophones.

At 4 a.m. on the 8th lookouts aboard the *Wolverine* made out an oil slick on the surface, and depth-charged it. Below the surface there was a great red flash and a tremendous explosion. The Bull of Scapa had received the *estocada*.

Kretschmer's *U99* and Joachim Schepke's *U100* picked off several convoy stragglers in heavy seas, and early on the morning of 16th March were ordered to join Fritz Lemp who had sighted inward-bound HX112 and was shadowing it.

Just before midnight on 15th March *U100* fired a fan of four torpedoes from about a mile outside the starboard escort screen, and hit the 10,000-ton tanker *Erodona*, which blew up.

Kretschmer slid through the escort screen into his favourite position between the columns of ships, fired all his remaining torpedoes and sank five ships.

Covering HX112 was the escort group under Commander Donald MacIntyre RN, comprising the destroyers, *Walker, Vanoc, Volunteer, Sardonyx* and *Scimitar*, and the corvettes *Bluebell* and *Hydrangea*.

MacIntyre was new as an escort commander, but he kept the convoy firmly together, and maintained a systematic and aggressive search round it. His own ship *Walker* dropped DCs over one contact, then his thoroughness was rewarded when *Vanoc* sighted a surfaced U-boat.

Joachim Schepke, on the bridge of *U100*, saw the destroyer heading towards him, but in the bad light mistook her angle of approach. He turned aside and said to the others on the conning tower, 'It's all right, she's going to miss us astern.' Then the handsome, dashing Schepke, cap at its usual jaunty angle, turned back towards the enemy, and was crushed between bridge coaming and periscope standard as *Vanoc*'s towering bows bit into the doomed U-boat.

MacIntyre in the *Walker* had just joined the *Vanoc* when the latter signalled 'U-boat surfaced astern of me.' Right over Lousy Bank *U99* came up, badly damaged by the *Walker*'s earlier attack, and was caught in the *Vanoc*'s blindng searchlight. Both destroyers opened fire, then English-speaking Kretschmer flashed, 'We are sinking.' Soon he was climbing up over the *Walker*'s rail.

A copy of a British press photograph showing Kretschmer striding down the gangway at Liverpool mollified to some extent Dönitz' regret at the losses. The deaths of Prien and Schepke were especially keenly felt, but he saw no point in withholding the news from the German public. However, High Command would not agree, and did not announce the loss of Schepke and Kretschmer until the end of April, and the sinking of *U47* three weeks later.

In Bordeaux, as a result of continued lobbying by Göring, control of KG40 was restored to the Luftwaffe, and the veteran anti-shipping pilot Oberst Martin Harlinghausen was sent there as Fliegerführer

Atlantik. Fortunately, he and Dönitz co-operated well.

On 9th May Fritz Lemp in *U110* attacked OB318 off the Hebrides. He missed with three torpedoes, and a fourth aimed at a big 15,000-ton whaling factory ship misfired. He was so annoyed with himself and so keen to have another shot at the whaler that he did not notice the *Aubretia* making a dead set at him. *U110* crash-dived but the corvette's DCs badly damaged her.

The destroyers *Bulldog* and *Broadway* (ex-USS *Hunt*) joined in a DC barrage, which wrecked the sub's rudder and hydrophones, knocked out the electric motor and cracked the battery cells. Deadly green chlorine gas leaked out. She surfaced in a hurricane of shot, shell and DCs. Lemp saw the *Bulldog* coming in fast, thought she was about to ram, and jumped into the water; but the destroyer swerved aside and lowered a boat, and Lemp, swimming doggedly, realised with horror that *U110* might be captured. He turned round and tried desperately to get back to the submarine. He had reached *U110*'s slippery hull and was trying to claw his way up when an armed seaman in the *Bulldog*'s boarding party took aim with a rifle and shot him dead.

U110 was taken in tow, and the treasures found below brought aboard the *Bulldog*. There were confidential code books and papers, but most precious of all was the U-boat's Enigma coding-decoding machine, together with sets of spare rotor wheels, the current daily rotor setting, diary and patrol signals. The capture of this machine, together with coding documents captured from the weather ship *Munchen*, enabled the cryptanalysts of the British Government Code and Cypher School in their mansion at Bletchley Park in Buckinghamshire to crack the Kriegsmarine M-Home Waters, or Hydra, code used for radio communication with U-boats. It was frustrating for Baker-Creswell of *Bulldog* that his coup was marred when *U110* sank as he was bringing her in.

It was an even more serious loss for Dönitz, though some of his other boats helped to even the score. On 1st June one of them torpedoed and sank the first of the thirty-five new Catapult Aircraft Merchant, or CAM ships, the SS *Michael E*, as she was steaming to join a westbound convoy. The Hurricane on her catapult was also lost.

In June also Oesten's *U106* berthed alongside the depot ship *Ysère* at Lorient after a sortie to West African waters lasting four months, during which she and Schewe's *U105* had been replenished in Spanish ports and by the supply ship *Nordmark*. The two boats had intercepted the slow convoy SL68, heading for England from Freetown, and from the Cape Verdes to the Canaries had played cat and mouse with it and its ponderous protector, the battleship *Malaya*. They came home having sunk seven ships from the convoy and damaged several more, but they did not know until Dönitz himself told them that one of Oesten's last two torpedoes had hit the *Malaya*, which did not sink but was out of action in a New York yard for several months. In the months of April and May the Allies lost 142 ships, a total tonnage of 818,000 tons, most of these in convoy actions. They were losses that besieged Britain, depending for her life's blood on what her ships could bring into port, could not sustain. Three merchant ships were being sunk for every new one launched, whereas, after the disasters of March, the average monthly production of new U-boats was back to seventeen or eighteen, for losses of between one and four. The agreement in April by the USA to cover convoys within five hundred miles of the American coast did not help the ships in the mid-Atlantic gap, where there was no air cover.

In August occurred a bizarre and untypical incident. *U570*, with a captain making his first patrol in command, surfaced eight miles south of Iceland and was bombed by a British aircraft. Her new captain panicked and waved his white shirt in surrender. Next day the Royal Navy captured the boat, she was taken into port, and later commissioned as HMS *Graph*. This balanced the capture by the Germans in May 1940 of HMS *Seal*, later renamed *UB* and used in the Atlantic.

On 4th September, in response to an order by President Roosevelt to the US Navy to attack all U-boats, Fraatz's *U652* was depth-charged by the US destroyer *Greer* 180 miles south-west of Reykjavik, Iceland. This was war in all but name, but Hitler still refused to allow his U-boats to attack any warships not clearly identified as hostile.

On 13th September a new element was added to the Battle of the Atlantic when the new auxiliary aircraft carrier HMS *Audacity* joined the escort for Convoy OG74 to Gibraltar. Foiled by the re-routing of convoys further north and west, it was on this route that the Condors were now concentrating. The primitive little carrier had no hangar, and just six American Wildcat fighters, renamed Martlet by the Fleet Air Arm of the Navy, whose Number 802 Squadron manned them. The pilots were pleasantly surprised to find themselves assigned to the ex-cargo liner's original staterooms. She had no bridge island, only a simple steel box offset to starboard level with the flight deck.

The *Audacity*'s main brief was to attack Condors, but from the beginning her patrols were used to look for U-boats. If a U-boat was sighted on the surface a Martlet dived and attacked with its machine-guns to force it down; then it climbed above it so that the carrier could get an accurate fix with her RDF to pass to the surface escorts. It was soon obvious that the aircraft needed something more lethal than ·50-calibre bullets to throw at a U-boat.

Two Martlets put down a U-boat on the 15th, greatly reducing her chances of catching the convoy. On the 20th the dusk patrol sighted a U-boat diving twelve miles west of *Audacity*, and from an RDF fix two escorts attacked and damaged the submarine, though two ships were sunk by another U-boat that night.

On 11th November OG76, the *Audacity*'s third convoy, reached Gibraltar, without loss. In Lorient Dönitz reported '. . . the worst feature was the presence of the aircraft carrier. Small, fast manoeuvrable aircraft circled the convoy continuously, so that when it was sighted the boats were repeatedly forced to submerge or withdraw. The presence of enemy aircraft also prevented any protracted shadowing or homing procedure by German aircraft. The sinking of the aircraft carrier is therefore of particular importance not only in this case but also in every future convoy action . . .'

With only one aircraft left serviceable, the *Audacity* remained at Gibraltar for a month while the rest were repaired. Sinkings by U-boats, which had reached a frightening peak in April with the loss of nearly 700,000 tons, had declined to 80,310 tons (twenty-three ships) in August. Then a surge in U-boat production made itself felt, and Allied losses began to rise again in September, only to fall off again. In November thirteen ships, a total of 104,640 tons, were sunk. By early December there were only some fifteen U-boats in the North Atlantic. But eighteen others had slipped through into the Mediterranean to support the German army in North Africa. While the *Audacity* was still at Gibraltar Guggenberger's *U81* sank the carrier *Ark Royal*.

On 7th December the Japanese attacked Pearl Harbour. Four days later the Germans declared war on the United States. American warships which since early September had been escorting convoys as far as a 'Mid-Ocean Meeting Point' in the Atlantic were now fully committed to the anti-U-boat war, though the US Navy would also be short of ships for some time, with its commitments in two oceans.

After two postponements, the *Audacity* sailed from Gibraltar on 14th December to give cover to Convoy HG76.

Dönitz had organised a strong ambush for the convoy. A special U-boat group operating under the name *Seerauber* (Pirates) was to concentrate round the convoy. This group included among others Baumann in *U131*, Gengelbach in *U574*, Hansmann in *U127*, and Muller in *U67* and other boats were alerted, including Heyda's *U434*, *Bigalk's U751* and Engelbert Endrass's *U547*. KG40 was standing by for a maximum effort. All were ordered to single out the aircraft carrier for special attention.

The Admiralty had anticipated a particularly heavy onslaught on HG76, and its escort was the strongest that could be provided for any convoy at this time. Thirty-two merchantmen, disposed in nine columns under Vice-Admiral Sir R. Fitzmaurice as Commodore, were flanked by nineteen escorts, including the *Audacity* and the 36th Escort Group under Commander F.J. Walker, who was making his first trip as a convoy escort commander. Between them the ships were carrying every weapon hitherto devised to deal with attacking aircraft and submarines.

The convoy was to have air cover from Gibraltar for the first three days of passage. The very first evening out a Sunderland sighted a U-boat. The night was black and moonless. Soon after first light next morning the Royal Australian Navy destroyer *Nestor* picked up a ping on her asdic, and about eleven o'clock her DCs destroyed Hansmann's *U127* with all hands.

At 4.47 a.m. on the 17th Baumann's *U131* reported, 'Contact established' to U-boat Headquarters.

The other U-boats were ordered to converge on Baumann's position, and at 6.05 Muller in *U67* sighted Baumann. At 7.45 Baumann reported, 'Contact lost', and a Condor was ordered to find the convoy and shadow it.

Now the convoy was nearing the extreme range of RAF aircraft from Gibraltar, and the *Audacity* began flying regular patrols. Just after 9 a.m., when the fuel of the two Martlets on the dawn patrol was running low, one of them reported, 'U-boat on surface twenty miles on convoy's port beam.'

This was Baumann's *U131*. The Martlet dived on him, the U-boat crash-dived, and the fighter circled round the spot for the *Audacity*'s RDF to get a fix and direct Walker's escorts to the scene. A detachment of the 36th reached the area and began a hunt.

Some of the other U-boats were approaching the convoy. At 10.01

Muller sighted smoke, and at 10.47 a boat reported the position of the convoy. At 10.50 a.m. the corvette *Penstemon* got a ping and dropped a DC pattern. There was no sign of a hit. The explosions were heard in *U67*.

At 12.47 p.m. the destroyer *Stanley*, ex-USS *McCalla*, signalled, 'Object on horizon to starboard,' and a few minutes later, 'Object is U-boat.'

It was Baumann again. Walker's ships went to full speed and converged on him, and Walker ordered Sub-Lieutenant Fletcher, flying one of the Martlets of Black Section, which had relieved the dawn patrol, to attack the U-boat in the hope that the submarine would engage the fighter with her guns and give the escorts more time to reach the scene.

Fletcher flew across the convoy with throttle wide open, saw the U-boat and dived on her.

As Walker had hoped, the U-boat did not dive, but opened fire on the fighter with her guns. As Fletcher pressed his gun button the glowing tracers began to float up towards the Martlet. His four Brownings began to yammer just as a cannon shell shattered his windscreen. The fighter jinked out of its dive and crashed smoking into the sea very close to *U131*.

The *Stanley* and the sloops *Stork*, *Blankney* and *Exmoor* steamed at full speed towards the U-boat, which had not dived. Leutnant Wiebe in Muller's *U67*, which had sighted *U131* again, recorded the leading hunters at 'Three smoke clouds' heading for the U-boat. A few minutes later a flying boat appeared and dropped bombs near *U131*, then circled and repeated the attack. Muller heard Baumann report, 'Ship unseaworthy. Am being chased by four destroyers.' Walker's ships opened fire on him from long range.

Twenty minutes later the *Stork*'s masthead lookout reported, 'Enemy abandoning ship. Looks as though she's been badly hit.'

The submarine's crew began jumping off the conning tower and casing into the rough sea. Baumann was picked up and sympathetically treated aboard HMS *Stork*, Walker's ship. Fifty-five survivors of *U131* were rescued. One of them said that the sub had received eight direct hits from shellfire, but was not sure about the effects of the *Penstemon*'s DC attack. Another man revealed that this boat had been acting as the shadower, and claimed that she had spent the whole of the previous night inside the convoy restricting herself temporarily to homing other U-boats in for the kill.

The *Stork*'s whaler brought in young Fletcher's body from the floating wreck of his Martlet and headed back towards the convoy. A Catalina forced Muller to crash-dive. *U67* surfaced again an hour later but almost immediately another flying boat appeared and put them down again. They stayed down this time for nearly two hours, surfacing again just in time to hear Gelhaus report contact with the convoy. They altered course accordingly, but twenty minutes later aircraft forced them to dive again in a hurry. They surfaced just over an hour later and heard Lorient ordering Baumann and Hansmann to report their positons. There was no reply.

At 9.06 a.m. next day, 18th December, the *Stanley* reported a U-boat contact six miles on the convoy's port quarter. Three sloops, led by the *Stork*, began a determined asdic hunt, picked up the submarine and saturated the sea with DCs. After half an hour of this battering the grey shape of Heyda's *U434* rose up, water cascading off her casing, men already jumping out of her conning tower hatch. Heyda joined Baumann aboard the *Stork*.

The *Audacity* was now reduced to three airworthy fighters. Just after the dusk patrol had returned that evening, the *Penstemon* reported another U-boat on the surface nine miles off the convoy's port beam. Three escorts hunted fruitlessly through the night, and were just rejoining the convoy at the end of the graveyard watch in the early hours of the morning when Gengelbach in *U574* got HMS *Stanley*, the leading escort, in his sights and fired. The old but very active American destroyer blew up in a great sheet of flame.

Walker immediately turned the *Stork* towards the burning wreck of the *Stanley*, got a ping and began depth-charging. After the fifteenth charge the *U574* broke surface two hundred yards ahead of the sloop. The *Stork* made full speed to ram. Gengelbach tried to escape by putting his helm hard over and turning inside *Stork*. He was so close that the sloop's 4-inch guns could not be depressed enough to bear on her. *Stork*'s First Lieutenant manned a stripped Lewis gun on the bridge and raked the U-boat's conning tower. After three circuits *Stork* rammed *U574* just for'ard of her conning tower, and dropped ten shallow-set DCs which completed her destruction. Only sixteen of her crew were rescued.

The *Stork* had damaged her bows and asdic dome but managed to pick up survivors from the *Stanley*. There were only twenty-five. The corvette *Samphire* rescued the German survivors. An hour after *U574* had sunk, *Audacity* was narrowly missed by a torpedo, and the leading ship in the port wing column, SS *Ruckinge*, was hit by another. She was abandoned, and *Samphire* sank her with gunfire.

Now Condors began to sniff round the convoy. Sub-Lieutenant Brown shot down one of a pair that morning, Sub-Lieutenant Lamb damaged the other. Sub-Lieutenant Sleigh destroyed another in the afternoon in a dangerous head-on attack which left him with part of the Focke-Wulf's W/T aerial wrapped round his tail wheel and a hole in his fuselage further forward.

At 3 p.m. next day, 20th December, a Martlet patrol reported two U-boats lying in wait dead ahead of the convoy, which then altered course fifty degrees to starboard to avoid them. Martlets kept any U-boats near the convoy submerged while the light lasted.

At 9.10 next morning Red Section were orbiting twenty-five miles astern of the convoy when they sighted two U-boats lying alongside one another, with a plank between them and men moving across it. One of the submarine had a hole in her port bow. As he approached they opened fire on him with cannon.

Brown positioned his Martlet right over the two boats, where their guns could not bear on him, and dived steeply, shooting three men off the plank and forcing the rest to scramble below. The two U-boats made off slowly on the surface, but *Audacity*'s RDF had been tracking the Martlets and gave the escort sloops a fix. Walker sent four escorts under *Deptford* to give chase. An hour later Brown reported them about twelve miles from their quarry, but then had to return to the carrier. His relief picked them up on the convoy's port beam. Walker sent the *Marigold* and *Convolvulus* to attack them; then he sighted another fine on his port bow, about ten miles away. At 3 p.m. he saw another, twelve miles off. 'The net of U-boats round us,' he reported, 'seemed at this stage to be growing uncomfortably close in spite of *Audacity*'s heroic efforts to keep them at arm's length.' He decided to wait for nightfall, then make a drastic alteration of course.

It was dark when the dusk patrol landed aboard the *Audacity*, and they only made it with difficulty, and the now standard use of hand torches by the 'batsman', Sub-Lieutenant Patterson.

As soon as they were down the *Audacity* turned to leave the shelter of the escort screen and begin her night zigzagging clear of the convoy, as she always did. This time, however, no escort could be spared to go with her. Walker did not agree with the decision to leave the convoy, but could not countermand it because Commander McKendrick of *Audacity* was senior to him.

The convoy made the planned alteration of course to throw the U-boats off the scent, and Walker sent *Deptford*'s striking force to stage a mock battle with depth-charges and starshell astern of the convoy to decoy the enemy further away. But some of the merchant captains, who did not know of the plan, thought it was the real thing, and fired snowflake, illuminating the whole scene for the benefit of the lurking remnants of the wolf pack.

At 8.33 p.m. the SS *Annavore*, rear ship in the centre column, was torpedoed. Another burst of snowflake lit up the sky, and Oberleutnant Bigalk on the bridge of *U751* saw quite clearly the unmistakeable silhouette of the aircraft carrier within torpedo range.

Four minutes later, while *Audacity*'s officers were finishing dinner in the wardroom, a torpedo struck the ship with devastating force. Plates and coffee cups smashed on the deck. The ship had a merchantman's thin skin, and when they got out on deck she had already settled well down by the stern, with her engine room flooded. She still had some way on her, but her rudder was jammed and she could not be steered. McKendrick ordered the engines stopped to avoid a collision. The *Audacity* was then dead in the water.

Bigalk's torpedo party were working frantically to reload as the U-boat closed in.

About twenty minutes after the first hit, men aboard *Audacity* saw the U-boat about two hundred yards on the carrier's port beam, glowing weirdly in the darkness with St Elmo's Fire. *Audacity's* 4-inch gun under the after end of the flight deck was awash and could not be trained, but a seaman got off a few rounds from P2 Oerlikon; then the U-boat fired two torpedoes which hit the carrier well forward.

There was a tremendous explosion, probably of aviation petrol, which blew off the whole fore part of the ship. The remainder reared steeply into the air, with the wire lashings holding the aircraft down whining under the strain. The first one snapped with a loud twang, and all the others gave way. The tubby Martlets hurtled down the plunging deck, crushing men, crashing into each other, then fell into the sea, on to the mass of men who had jumped off the tilting deck.

At 10.10 p.m. what was left of *Audacity* sank, her single propeller high in the air. Swimmers felt the shock of underwater explosions.

Half an hour later *Deptford*, over on the port beam of the convoy, sighted a U-boat between her and the merchantmen. After a two-hour asdic/depth-charge attack the ace Engelbert Endrass, his boat *U547* and all his crew were destroyed. *Samphire* and *Marigold* attacked other contacts, and forty-one DCs almost put paid to Muller's *U67* six hundred miles north-west of Finisterre. When they finally surfaced they saw that they were trailing an oil slick twenty yards wide.

The survivors from *Audacity* were in the water for three hours before escorts could break off from the urgent hunt for U-boats to rescue them. By then many had died. Among those drowned was Commander McKendrick. In the forenoon a Coastal Command Liberator joined the convoy and escorted it for two and a half hours without any U-boat being sighted. Dönitz recorded: 'The chances of losses are greater than the prospects of success.' Just after dawn on 23rd December all U-boats in the area of HG76 were signalled, 'Break off operations against convoy and return to base,' On Christmas Day the escorts left the convoy, and thirty ships, out of thirty-two, came sailing in.

CHAPTER TEN

Shallow Waters

The Mediterranean is a shallow sea, and the O, P and R submarines arriving to form the Royal Navy's 1st Flotilla at Alexandria in April and May 1940, should not have been there at all. They had been built in the late twenties, when Japan was the potential enemy, designed to make the long ocean passage to the Far East and operate in the deep waters there. At over 2,000 tons, 260 feet in length, they had a silhouette as big as a destroyer's when surfaced, and they were bulky targets underwater for Italian asdics. But they were all the Admiralty could call on at the time to sink transports plying between Italy and her North African colonies. With six tubes and a 6-inch gun apiece, they might yet do some damage, though crews were tired from two years in the tropics, with no home leave in sight.

When war ended in 1918 the Royal Navy cut its submarine strength from 138 boats to 55, phased out the veteran E, H and most of the L classes, scrapped nearly all the big coal-burning K boats, and, except for the monster submarine cruiser *X1*, with its 3,600 tons and four 5·5-inch guns, built no more new boats until 1926. Three of the K class were converted, two into submarine monitors with one 12-inch gun apiece, one to carry a small Parnell Peto seaplane in a hangar. Their unusual features, like the disappearing funnels of the original K boats, and the special drill essential for diving, made these freaks potentially dangerous. *M1* was lost in an accident off Start Point in 1925, *M2* and her aircraft off Portland in 1932. *X1*, which was as reminiscent of the big U-cruisers as were the British airships *R100* and *R101* of the German Zeppelins, was scrapped in 1931.

Between 1926 and 1930 the nineteen boats of the O, P and R classes were built, and between 1932 and 1935 the larger *Thames*, *Severn* and *Clyde* Fleet boats, with their 4·7-inch gun and 22½ knots top surface speed. In the late thirties, with Germany once more becoming a threat, there was a reversion to smaller boats, more suitable for home and coastal waters. The S class was the first of the new trend, displacing 715 tons surfaced, 990 tons submerged, followed by the 1,090/1,575-ton T and the smaller, handier 540/740-ton U class.

In April 1940 the fleet returned to Alex. The *Warspite* steamed in. Two other battleships, *Malaya* and *Royal Sovereign*, joined her from the Atlantic, *Ramillies* came in from Eastern waters, bringing with her the aircraft carrier *Eagle*. Six cruisers swelled the numbers. They look impressive, but Cunningham's fleet was heavily outnumbered by Mussolini's modern armada.

By the end of May the submarine depot ship *Medway*'s clustering brood comprised *Odin, Olympus, Orpheus, Otus, Pandora, Parthian, Grampus* and *Rorqual*, though *Olympus* and *Otus* were booked for a refit in the dockyard at Malta, and on their way from the Far East, where for ten years they too had sweated and strained in typhoons and tropical heat, were *Perseus, Rainbow, Regent* and *Regulus*, their bones creaking from advancing age, their hydroplanes temperamental.

On her first war patrol HMS *Parthian* was steering west along the North African coast. Then at midnight she received a radio message that an enemy submarine had been sighted on her way to Tobruk. The *Parthian* retraced her course in the darkness, diving before dawn and taking up station across the enemy sub's likely path.

The morning watch crawled by, and hands were piped to dinner. The non-duty watch were still eating at one o'clock when the watch-keeping officer raised the periscope to make another search.

'Tell the Captain there is a long low object dead ahead.'

Lieutenant-Commander Rimington took over the periscope. There was the target, exactly where he had calculated she would be. Four torpedoes whooshed out of the tubes at three-second intervals. They counted the seconds. There was a distant crash, then another, and another. In the control room they were grinning and looking at the unruffled Rimington, when there came the shock of a fourth hit. It was a record. Then the whole boat reeled to the impact of yet another explosion, greater than all the others, signalling the final disintegration of their target, the Italian submarine *Diamante*. It was the first sinking of an enemy vessel in the Mediterranean in World War 2.

Handling a submarine in Mare Nostrum called for particular skill, and unfortunately some of the submarine COs had not had Rimington's experience in command. When *Perseus* joined the flotilla in the second week in August her ship's company were appalled to find that no fewer than five boats, *Odin, Orpheus, Oswald, Phoenix* and *Grampus*, had failed to return from patrol, without a single success reported.

Then Hugh Haggard in the newly arrived *Truant* sank the 3,000-

...erleutnant S. Rahn
...ntre) with officers from
...3.

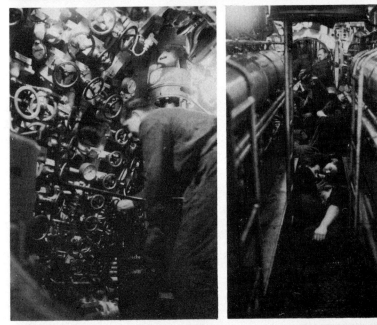

...de a U-boat.

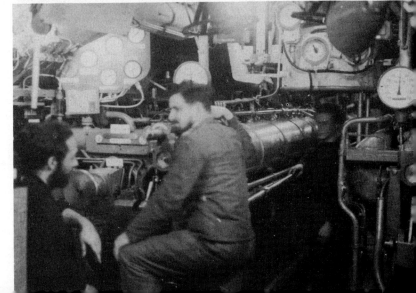

Grumman Avenger TBF1.

U288 under attack by Avenger from HMS *Tracker* (*Top left*) The first Avenger attack from the lower aft position. (*Top right*) The second attack. (*Bottom left*) The third attack seen from the Avenger's bomb bay. (*Bottom right*) The fourth attack.

ton supply ship *Providenza* in the Bay of Naples. On 22nd September Harvey in the old *Osiris*, patrolling in the Adriatic off Brindisi, attacked the destroyer *Palestro*. One kipper hit her in the bows, and smoke started to billow out; then a much greater explosion, possibly her boilers going up, blew out her side, and she sank. Back in Alexandria Harvey was presented with the first submarine Jolly Roger, a black flag with a white skull and crossbones, afterwards flown by all British Mediterranean submarines, white bars being added for ships sunk, stars for gun actions, daggers for landings of spies and saboteurs.

Watkins of *Triton* put up a bar for the 5,000-ton ship sunk off Genoa, two stars for the factory bombarded at Savonna and the gasworks gunned at Vado, but was reported lost on a patrol in the Adriatic. Currie's old *Regulus* and Salt's new *Triad* were both sunk by mines. In December Haggard in the *Truant* sank a 1,500-ton ship off Calabria, and capped this with the bigger prize of the 8,000-ton tanker *Bonzo*. By that time the Fleet Air Arm had reduced the number of Italian warship targets by knocking out three of Italy's six battleships and two cruisers in Taranto harbour with a squadron of Stringbags.

The record of British submarines in the Mediterranean for the first six months of the war against Italy was discouraging. Nine boats had been lost for the score of only six supply ships and one tanker, 29,000 tons' worth in all, known to have been sunk, plus *Parthian's* submarine and *Osiris'* destroyer, and the Admiralty now set up a submarine force in Malta, from which only a few Os and Ps had been operating sporadically so far, with new T and U class boats. The latter would be much more suitable for the shallow waters off Libya, though they could only make nine knots underwater, twelve on the surface, and were not fast divers.

Towards the end of 1940 the first five U class submarines, Lieutenant A.J. Mackenzie's *Ursula, Utmost*, which was taken over by Dick Cayley, Lieutenant E.D. Norman's *Upright*, Lieutenant A.F. Collett's *Unique*, and Lieutenant-Commander W.D. Wanklyn's *Upholder*, arrived at the submarine base at Lazaretto on Marsa Muscetto Bay, just north of the capital Valletta, on the south-east coast of the island, to operate under the command of Captain C.W.G. Simpson.

These new boats were at first plagued by engine troubles, and were also held back by a shortage of torpedoes in Malta. New ones had to be brought in small numbers by submarines arriving from

Gibraltar. Otherwise old destroyer kippers had to be converted, and these were undependable.

The tall, reserved, thirty-year old David Wanklyn, captain of HMS *Upholder*, married, with a child, was an experienced submarine officer. There was no naval tradition in his family, and it was a walk round a submarine when he was a small boy in the First World War that had kindled the ambition to captain a submarine. He was high in his class in Dartmouth, and joined his first submarine, *Oberon*, in 1933, proceeding to the old *L56*, and the even older *H50*, the new *Sealion*, HMS *Shark* as First Lieutenant, and *Porpoise*, under 'Shrimp' Simpson, who was now his Captain (S) at Malta. After *Otway* as First Lieutenant, he was given his first command in August 1939, serving in *H31* and *H32*, both boats completed in 1919 as copies of a wartime US-built type in British service, which provided asdic targets for surface ships off Portland. Submarines were scarce, and he was sent on hostile patrols in the old *H31* from Blyth, Northumberland, before taking command of *Upholder* in August 1940.

On the night of 27th January 1941 *Upholder* was surfaced near Kerkeneh Bank off the Tunisian coast on her first war patrol when she sighted two big ships, one an armed merchant cruiser, the other a motorship. Wanklyn opened his score by hitting the 7,889-ton German *Duisberg* and putting her in dock in Tripoli for four months. Three days later he sank a 3,950-ton supply ship.

Early in February Hitler summoned General Erwin Rommel to the Chancellery and gave him command of the new Afrika Korps which he was sending to North Africa to stop the British from taking Tripoli and drive them back through the desert. On 8th February the first convoy of troops and supplies left Naples, the German freighters *Ankara*, *Arcturus* and *Alicante*, with destroyer escort. Meanwhile Swordfish aircraft from Malta laid mines off Tripoli.

On 12th February off the Gulf of Gabes on the Tunisian coast Pat Norman's submarine HMS *Upright* hit an 8,000-ton supply ship with three torpedoes. Norman then patrolled off Kerkeneh Bank. Mackenzie's *Ursula* was in the same area, and on the morning of the 22nd scored torpedo hits on the 2,365-ton *Silvia Tripcovitch* and the 5,788-ton *Sabbia*. Neither ship sank, but that night *Upright* sighted the damaged *Silvia Tripcovitch* south-east of Sfax. Her torpedo hit the valuable tanker and she disintegrated in a violent flash and sheet of flame.

The *Upright* continued her successful patrol. In the dark of the graveyard watch early on 25th February she was lying on the surface charging batteries when the officer of the watch made out three ships, steaming fast. Norman fired a salvo in the blackness, then crash-dived in a hurry. Although he had sunk the 5,000-ton light cruiser *Armando Diaz*, he was kept down by depth-charging long enough to miss a chance of attacking the big convoy of four loaded troopships, the former liners *Esperia, Conte Rosso, Marco Polo* and *Victoria*, which the *Diaz* and her screen were protecting.

With Number 148 Squadron's Wellingtons at Luqa almost destroyed by bombing, only a depleted Swordfish unit and the handful of submarines were left in Malta to attack the Libyan convoys, which were getting through regularly to Tripoli with troops and supplies for the Axis build-up in Cyrenaica prior to an attack on Wavell's army, which had been seriously reduced by detachments sent to Greece to face a German drive there.

The parched and weathered Stringbags took off on nightly 'rat hunts' to Kuriat and Kerkeneh off Tunisia, often in bad weather, or stood by at bomb-battered Hal Far in case a recce Maryland of Number 431 Flight or a Sunderland of 288 Squadron reported a target. But the recce aircraft were few, and could not be everywhere.

Between the 8th and 12th of March two German supply convoys reached Tripoli safely, but Cayley and Collett got amongst the five supply ships of an Italian convoy en route from Trapani to Tripoli. Collett in *Unique* sank the *Fenicia* of 2,584 tons, Cayley in *Utmost* destroyed the 5,683-ton troopship *Kapo Vita* in the Gulf of Hammamet, and later sank the German *Herakleo* of 2,000 tons off Kerkeneh. The three troopships *Conte Rosso, Marco Polo* and *Victoria* sailed from Naples on the 12th, reaching Tripoli without loss, and another supply convoy for the Afrika Korps unloaded at Tripoli and returned to Naples undisturbed.

Just before two o'clock on the afternoon of 15th April a Malta Maryland reported a slow southbound convoy half-way between Pantellaria and the Gulf of Hammamet, making about eight knots. At 6 p.m. the four destroyers *Jervis, Janus, Mohawk* and *Nubian* of Captain Mack's 14th Flotilla sailed from Grand Harbour, Valletta, under cover of rain and low cloud.

They intercepted the convoy between Numbers 3 and 4 Buoys off Kerkeneh Bank. There were five medium-sized merchant ships escorted by three destroyers. One Italian destroyer and four of the merchantmen, the German *Adana, Iserlohn* and *Aegina*, all taking troops and supplies to Rommel, and the Italian ammunition ship *Sabaudia*, were sunk outright, and the 2,452-ton German trooper *Arta* and a damaged Italian destroyer beached on Kerkeneh Bank.

On 21st April *Upholder* left Malta for her billet off Kerkeneh with orders to finish off the *Arta* and the destroyer which had been reported still stuck on the Bank. Wanklyn had now grown a long black beard. On his way there, he sighted a merchantman, dived, and from a range of half a mile fired two torpedoes. Fifty seconds later the heavily loaded 5,482-ton motorship *Antonietta Lauro* went down.

Next day *Upholder* approached Kerkeneh Bank, located the grounded *Arta* and boarded her to set explosive charges. The destroyer was found too difficult to approach, but the fire which raged in the *Arta* when the charges were set off lit the whole of Kerkeneh Bank.

On May Day Wanklyn sighted a convoy of five transports and four destroyers. *Upholder* strained forward at her best nine knots, with Wanklyn as always wishing for more speed in his otherwise agile boat. At 11.45 a.m. he opened fire with a relay of four kippers, sank the 2,586-ton German motorship *Arcturus*, one of the first ships to take Afrika Korps troops to Tripoli, outright, crippled the bigger, 7,836-ton *Leverkusen*, went deep, and then returned to finish her off. *Upholder* berthed at Lazaretto on 3rd May with some of her crew on the casing wearing German helmets taken from the *Arta*. Many Maltese watching her thought she was a U-boat coming in to surrender, and the rumour flashed quickly round Sliema and Valletta.

Wanklyn's latest victories, duly recorded on his Jolly Roger, were particularly important as Rommel was at Sollum, just inside Egypt, building up his forces for a new attack. Tanks, men and supplies were arriving at Tripoli and Benghazi for his panzer divisions, challenged only by a few submarines, RAF Blenheims and Wellingtons and Naval Swordfish in Egypt and Malta.

Upholder left Malta for her seventh patrol on 15th May, heading for the Straits of Messina. She fired three torpedoes at a convoy, then endured a long and frightening battering as depth-charges burst close around her. Glass and loose gear littered the confined spaces, asdics and hydrophones were knocked out. But she escaped, and three days later hit the tanker *C. Damiani*, which was towed, badly damaged into Messina.

At 8.30 p.m. on 24th May *Upholder* was on the surface charging batteries when three big troopships were sighted and attacked. The 17,897-ton former Lloyd Triestino liner *Conte Rosso*, launched in

Britain in 1923, already a veteran of the Libya run, went down fast. Of the 3,000 troops aboard her, only 1,432 were saved by the escorts.

British submarines sank 89,797 tons out of the total of 209,714 tons of enemy shipping sunk in the Mediterranean in the period January to the end of May 1941. After *Upholder*, Tomkinson's *Urge* was the most effective boat. Before operating from Malta he had already sunk a 10,800-ton tanker. On his first Malta patrol, between Pantellaria and Lampedusa, he sighted a convoy. Penetrating the screen of five destroyers, he hit and sank the 5,000-ton *Zeffiro* and badly damaged the *Perseo* of similar tonnage.

For some months some of the old but roomy 'China' boats had been making the 'Magic Carpet' run from Alexandria to Malta with essential supplies. *Cachalot*, *Parthian*, *Regent* and *Rorqual* had all been specially adapted for this errand of mercy. One section of the batteries was removed to carry cargo, and some freshwater, fuel, even main ballast tanks were used for petrol, though the boats remained equipped for action.

On one of these trips the *Parthian*'s tanks carried aviation gas for the Hurricanes and Blenheims at Takali and the Stringbags at Hal Far, and her reload of torpedoes had been disembarked to make way for aircraft spares in the torpedo space. When these were packed in, a false deck was laid on top of them to create a mess for the soldiers and airmen they were transporting. The landsmen were very uncomfortable for much of the passage, but Rimington kept the boat on the surface for long as possible and allowed them, two at a time, to stand under the open conning tower hatchway and get some fresh air. By the time *Parthian* reached Malta they had recovered sufficiently to have eaten all the food in the boat.

In May the Germans drove the British forces out of Greece and into Crete, then, on the 30th the island too was finally evacuated. The capture of the Dodecanese Islands followed, and new patrol areas were opened for the 1st Flotilla boats at Alex, in particular the Aegean Sea, across which tankers steamed with Rumanian oil for occupied Greece and Italy.

The *Parthian* left Alex on 23rd May and was in her billet in the Dardanelles area six days later. On 3rd June she sighted an eastbound convoy of three tankers and one escort, and attacked and crippled the 5,000-ton *Strombo*, which had to be beached. *Parthian* was shaken by a persistent DC hunt, and her main engines damaged, but she escaped and made repairs. Rimington was then ordered to investigate shipping in the harbour of Mytilene in the island of

Lesbos. Very early on the morning of 8th June he conned the big submarine slowly into the harbour. It was a calm, windless morning, promising another blue and gold Aegean day. The sea was glassy, and a probing periscope would shatter the glass instantly.

Rimington slowed the boat right down, raised the lens slowly so as not to start a ripple, and made a short sweep round the anchorage. The biggest vessels in port were two big schooners and a dumb lighter. It was disappointing, but these were ships which would be useful to the enemy. In the open sea he would have surfaced and used the gun, but here it had to be torpedoes and a quick getaway, even though some of his kippers had been cured before the First World War. With the helm hard over, *Parthian* gently reversed course, and fired two thirty-one year old torpedoes into the harbour. One exploded amongst the boats, the other geriatric missile wobbled off course and torpedoed Lesbos, which remained afloat. It was, as no naval historian could possibly resist saying, a truly *Parthian* shot.

The old boat operated now very much on faith and familiarity, and her next patrol was to be her last before a badly needed refit in the USA. Unlike too many unlucky British subs she was able to make it a triumphant finale. With Greece in their hands, and Rommel practically on the Nile, the Germans were making moves to soften up the Middle East for a final coup which would give them a vast oilfield and a launching pad for a drive to link up with the Japanese in India. Their influence in Vichy-French Syria was growing very strong, and on 8th June Allied forces had gone into the country to stamp it out.

To support an infantry landing, the cruisers *Phoebe, Ajax* and *Coventry* and eight destroyers steamed off the coast. On 9th June the destroyer *Janus* was disabled by two French destroyers from Beirut, and on the 15th HMS *Isis* and *Ilex* were badly damaged by bombers. On the same day Royal Navy torpedo bombers from Cyprus retaliated by sinking the French flotilla leader *Chevalier Paul*.

Ten days after this, *Parthian* sighted the French cruiser submarine *Souffleur* at long range. She gave chase, but the Frenchman had seen her, and dived. Rimington continued to track the enemy, and when the submarine cruiser surfaced three hours later he did not miss the big target, and she sank.

By the end of June the tonnage of enemy ships sunk in the Med in 1941 by Royal Navy submarines had gone up to 122,225, or forty ships.

Against this on the scales were the losses of Lieutenant G.P. Darling's *Usk*, which did not return from a patrol off Marritimo, and

Lieutenant J.F. Livesay's *Undaunted*, which was sunk on her first patrol off Sicily.

Lieutenant Commander Miers had taken *Torbay* into the Aegean, and on the last day of June, just before 9 a.m. he sank a fully loaded caique with his gun, following this up next day by torpedoing the Italian tanker *Torsello* in the Zea Channel. Next morning he picked off one ship from a convoy of two motorships with an escort of one destroyer and an aircraft. On 4th July off Doro Island he sighted a caique and a schooner flying the Nazi ensign, loaded to the gunwhales with troops and stores. With *Torbay* at full buoyancy but held down by her planes, he suddenly rose up and sank them with his gun. Next day he switched to torpedoes again to dispose of the Italian submarine *Jantina*. On the 8th a trooping caique was gunned down off Cape Malea, and he topped off the patrol on the 9th when he boarded the two caiques and a big schooner in a convoy northbound from Crete, took off the caches of captured British ammunition, and sank the vessels, with their consignments of petrol and stores.

On the *Upholder*'s tenth patrol, having missed the first target with three torpedoes and been left standing by an Italian supply ship, a frustrated Wanklyn found a convoy of three medium-sized motorships on 3rd July and torpedoed the 6,000-ton *Laura C*. His next patrol also began with troubles, including near-destruction in a bad DC hunt and more faulty old torpedoes, but all this was forgotten when at half-past seven on the evening of Monday 28th July Wanklyn sighted two cruisers and two destroyers heading straight for his periscope. Eluding the zigzagging destroyers he fired a salvo, two torpedoes of which hit the bows of the cruiser *Garibaldi*, which stopped dead. He went deep, leaving her badly damaged, and was lucky to escape the thirty DCs which rained down on him.

The cryptographers at Bletchley Park had decoded an intercepted Italian signal which revealed that a fast convoy of four troopships, the *Esperia*, *Marco Polo*, *Neptunia* and *Oceania*, with seven destroyers and a torpedo-boat, were scheduled to leave Tripoli at 5 p.m. on 21st August, north-bound for Italy. *Unique*, temporarily commanded by Lieutenant A.R. Hezlet, sailed from Malta on 14th August with orders to form a patrol line off Tripoli with the newcomers *P32* (Lieutenant D.A. Abdy) and *P33* (Lieutenant W. Wilkinson).

On 18th August Hezlet was on station and called up the other two boats. *P32* answered but he could not raise *P33*. By the 20th *P32* was not answering either. Both boats had been sunk. Unable to get at their Davis Escape Apparatus, Abdy and two ratings of *P32* made an astonishing escape through the conning tower from 210 feet in a bubble of air.

At 10 a.m. that morning Hezlet sighted the four big troopers he was looking for, with a local escort of three flying boats. He managed to get in to 700 yards, and three of his four torpedoes sank the 11,700-ton *Esperia*, which had made several runs between Italy and Africa.

Early in September the Malta submarine force was formed into the 10th Flotilla, under Shrimp Simpson.

Upholder began her fourteenth patrol on 16th September. Air recce from Malta had reported three large liners in Taranto Harbour loading for another run, and had kept a regular watch on progress until the ships were reported to have left. *Upholder*, Hezlet's new command *Ursula*, Woodward's *Unbeaten* and Wraith's *Upright*, were ordered out to take station across their probable course along the convoy's alternative route east of Sicily via Misurata, eighty miles east of Tripoli.

The four boats were in position by 17th September, *Upholder* in the middle of the line, *Upright* five miles to the north, *Unbeaten* five miles to the south of her, and *Ursula* further west as a longstop. The convoy was probably expecting to reach Tripoli about 9 a.m. next morning, and the submarines hoped to intercept in the early hours.

Just after 3 a.m. *Upholder*, trimmed down with only her conning tower above the water, received a signal from *Unbeaten*, to the south-east of her, reporting the convoy sighted heading towards *Upholder's* billet.

Wanklyn headed to intercept at twelve knots, his night glasses searching the darkness. The boat was yawing badly in the choppy sea, but if he dived he would lose the convoy.

As the boat's bows swung across the line of shadowy ships Wanklyn fired four torpedoes. Four minutes from the firing of the first one the distant rumble of explosions began, as up above two old enemies of the Malta submarines were hit. *Neptunia*, 19,500 tons, was struck amidships and started to list. Her sister *Oceania*, holed aft and her props smashed, settled stern-down. Two destroyers steamed up to tow her. *Vulcania*, at 24,500 tons the biggest of the three, increased speed and fled with one destroyer for Tripoli. Wanklyn had no chance of catching her, but he came to periscope depth at 6.30 to find *Oceania* still afloat. Two torpedoes hit her, and she sank swiftly.

In this patrol alone Wanklyn had accounted for two-thirds of all the tonnage sunk by the British Mediterranean submarines in the month. At the end of the middle watch on 8th November he sighted an Italian submarine of the *Perla* class, fired four torpedoes and saw the target blow up. Just before he dived, with daybreak over the horizon, he received a signal warning him that Captain Agnew's Force K from Malta, with the cruisers *Aurora* and *Penelope* and the destroyers *Lance* and *Lively*, would be passing through his area that night. The submarine he had destroyed had been stationed to protect the convoy which Force K was now steaming to intercept. Heading south from the Straits of Messina were five freighters and two tankers, with an escort of six destroyers and a covering force of two cruisers and four destroyers.

At dawn on the 9th he watched three *Aviere* class destroyers searching the wreckage left by the seven supply ships, all of which Agnew had sunk. One of them, the 1,450-ton *Libeccio*, stopped, and Wanklyn got her in his sights. He had four torpedoes left, and wanted to save them for bigger game, but one was enough to leave *Libeccio* low in the water.

Up came cruisers *Trieste* and *Trento* of the covering force, their destroyers zigzagging on their flanks. Wanklyn had about ten minutes to get in an attack. His first torpedo grazed the bow of the nearest cruiser but blew the bows off the destroyer beyond her. The second torpedo went round in circles. The third fish also missed the cruiser by inches but hit the already damaged destroyer and sank her.

1942 began badly for 10th Flotilla, with *Triumph*, *Tempest* and *P38* failing to return from patrol, but *Upholder* on her twentieth patrol damaged a tanker, and sank the Italian submarine *Ammiraglio St Bon* with one torpedo, a second having exploded prematurely almost under *Upholder* herself.

After a long lull caused by the temporary withdrawal of Luftwaffe units to Russia heavy raids began again on Malta, and the submarines at Lazaretto suffered severely. *Una*, *Unbeaten* and *Upright* were damaged, *P31* had some thirty holes punched in her hull, and *P39*, towed to the dockyard with her batteries and engines useless, was wrecked by two direct hits, one of which broke her back. *P36* turned over and sank.

The big Italian submarines had suffered as much as the old British China boats in the shallow Italian Lake, and the attrition went on. Harrison's *Ultimatum* sank the *Ammiraglio Millo* close inshore off Cape Stilo, black-bearded Wanklyn destroyed the *Tricheo*, his third sub kill.

Unbeaten was repaired after the bomb damage she had sustained at Lazaretto and sank the submarine *Gugliemotti*. But in April the bombing of Malta reached a new peak of fury, with the airfields, the anchorages and the streets and houses of Valletta all taking a renewed battering. The block where the submarine officers lived was demolished, the submariners' rest camp bombed and machine-gunned, a Greek submarine refitting was destroyed.

The *Upholder* had done twenty-three patrols. For her last patrol before she went home for refit she was sent to the temporarily quieter waters off Sfax.

Wanklyn made his rendezvous with *Unbeaten* on 11th April then steered south for the western approaches to Tripoli, where *Upholder* and Tomkinson's *Urge* were to watch for two troopers which were reported about to leave port.

About 6 a.m. on the 16th Tomkinson heard the distant explosion of a single depth-charge. This was repeated on the hour throughout the forenoon, then these isolated warning shots were suddenly replaced by an outburst of continuous depth-charging, which did not cease until 6 p.m.

The noise of battle had been near the *Upholder*'s billet, but she did not reply to Tomkinson's signals. As hours, then days went by it became clear that there would be no more signals from *Upholder*. She would lie forever on the bottom of the shallow sea with the 82,000 tons' worth of ships she herself had sunk.

The bombing of Malta and the mining of the inshore waters had now grown so intense that Shrimp Simpson was forced to withdraw his submarines from the island. The five boats still operational left for Alexandria, but on 19th July Lieutenant Alastair Mars in *Unbroken* led *Unison*, *Unbending* and *United* into Marsa Muscetto to form a new Malta submarine force. In support of the big August Convoy Pedestal, which got five supply ships out of fourteen through to Malta, *Unbroken* attacked two cruisers, damaged the *Bolzano*, and blew the bows off the *Muzio Attendolo*.

Submarines harried the ships bringing replacements for the losses inflicted on Rommel at the battle of El Alamein on 23rd October, and eighteen boats, the largest number of British submarines ever deployed together, supported Torch, the Allied landings in North Africa.

Everywhere in the Central Mediterranean the submarines were part of the rising tide of Allied success. McGeogh's new *Splendid* sank the destroyer *Avierre* and the ammunition ship *Sarah Antiochia*.

Cayley's *P311* and the three 'chariots' she was carrying were lost en route for an attack on shipping in Maddelena and Palermo, but five other chariots carried by *Thunderbolt* and *Trooper* reached their targets in Palermo harbour. One sank the new light cruiser *Ulpio Traiano* with its detachable warhead, another badly damaged the 8,700-ton liner *Viminale*. One of the two-man crews was picked up out at sea by *Unruffled*. *Sahib* sank a motorship off the Côte D'Azur, bombarded a seaplane base at Finale, and sank *U301*. The bearded ex-Royal Navy Rugby forward Commander 'Tubby' Linton, with a destroyer and twenty-four merchantmen already in the bag, sank a whole convoy of four supply ships and a destroyer, before being lost off Corsica. Bryant's *Safari* attacked a heavily protected convoy in shallow water and sank an armed merchant cruiser, a tanker and a supply ship.

At the end of the African campaign the 8th Flotilla based at Gibraltar, and later Algiers, alone had sunk 250,000 tons of shipping in the six months since Alamein. Of 3,000 torpedoes expended, 900 had hit the mark. Another 10,000 tons had been destroyed by gunfire. Forty-seven Allied submarines covered Operation Husky, the invasion of Sicily, HMS *Shakespeare* provided a flashing beacon for the Salerno landings and sank the Italian submarine *Valella* just a few minutes before Italy officially surrendered to the Allies. The surrender terms included the handing over of Italy's merchant marine, and Lieutenant Turner took *Unrivalled* into Bari harbour, organised all the shipping there into a convoy, and escorted it to Malta.

All the Algiers boats left for a final fling in the Aegean. *Seraph* sank a big caique with a hand grenade, stopped a launch and captured the German Commandant of the area, torpedoed a loaded caique, gunned down another and an Arado seaplane, and bombarded a barracks. Wraith's *Trooper* and Milner's *Simoon* were lost in these blue waters. Right at the end of 1943, with the Mediterranean war folding up, Lieutenant Ketts RNR in *Ultimatum* sank *U431*.

CHAPTER ELEVEN

General Chase

The little *Audacity* had been a test case for seaborne air support of convoys. Although equipped only with aircraft powerless to sink submarines, her half a dozen Martlets had proved invaluable to the escort commander in locating them, and had contributed significantly to the six, possibly seven, U-boats destroyed in the battle for Convoy HG76, in which the carrier herself was lost.

Even before *Audacity* had been commissioned, both the Admiralty and the US Navy Board had become convinced of the potentiality of the 'auxiliary carrier', converted from a merchant hull, as a submarine killer. The US Navy's first of such ships, the *Long Island*, was commissioned a fortnight before *Audacity*, and in March 1941 the US Navy ordered the conversion of a batch of merchant hulls, of which the Royal Navy asked for six. Four American Fleet tankers were also taken in hand for conversion.

None of these ships was ready to combat *Paukenschlag* ('a bang on the kettledrum'), the onslaught by packs of U-boats on the stream of unprotected shipping steaming along the North American coasts in the early months of 1942. These sheep were slaughtered in masses until organised in convoys. The six American-built auxiliary carriers were all delivered by the summer of the year, but they were delayed, first by engine defects, then by 'anglicising' in British yards. This included the lengthening of their flight decks to allow the British Swordfish TSRs, which could not use the American catapults, to take off with a full warload, and the fitting of the HF/DF (High Frequency Direction Finding) or 'Huff-Duff', which located the position of a U-boat from intercepted radio signals. Then the carriers were used on Russian convoys and in Operation Torch, the Allied invasion of North Africa in the autumn of 1942, a time when the Allies were losing the Battle of the Atlantic once again.

U-boats were being sunk by a steadily increasing number of escorts, but German yards were producing new submarines twice as fast as losses. The worst attrition of Allied merchantmen was in the mid-Atlantic gap, where there was still virtually no air cover.

The Admiralty approved a scheme to convert twelve grain ships and tankers to Merchant Aircraft Carriers, to carry their original cargoes and three or four Swordfish anti-submarine aircraft. These were quicker to convert than auxiliary carriers, which were wholly warships, but would still take months to complete.

On 26th September 1942 USS *Bogue*, first of the new auxiliary carriers for the US Navy, was commissioned, and so great was the need for air cover in the North Atlantic that Admiral Ernie King, US Navy C-in-C, decided to lend her to the British to help them carry out their responsibility for that theatre of war. *Bogue* began trials and working-up with her new air squadron.

On 31st January 1943 Dönitz replaced Raeder as C-in-C of the German Navy, and inaugurated a new U-boat offensive. Boats were now being produced at a rate of twenty-seven a month, and as fast as these could be worked up they were sent on station in the stormy grey Atlantic.

The weather was particularly rough this winter, and was a mutual enemy to both convoys and their attackers. Gales scattered the sixty-one merchantmen and nine escorts of SC118 from Halifax, Nova Scotia, but Kapitänleutnant Münnich in the newly commissioned *U187* located this loose mass of ships, hung on for two days in the teeth of mountainous seas, and his homing signals brought twenty more boats to the attack.

Nine Allied ships were sunk, three U-boats lost, between 5th and 8th February.

In this mounting campaign of punch and counterpunch, the Allies began a new move to sink U-boats before they reached their operating areas. Most of them used the west-facing French ports of Brest, Lorient, St Nazaire, La Pallice and Bordeaux, crossing and re-crossing a patch of sea 300 miles by 120 in the Bay of Biscay.

Captain Norman Ives' US Navy Submarine Squadron 50 patrolled the southern sector from early in the year, handicapped by proximity to Luftwaffe bases, and by bad torpedoes, but the real offensive against these boats in transit through the Bay was started in February by Air Chief Marshal Joubert, commanding the RAF's Coastal Command. Working for him, the thirty-six Liberators of Colonel Jack Roberts' two anti-submarine squadrons of the US Army Air Force battled with the U-boats in Biscay, and Lieutenant Sanford's 'Tidewater Tillie' sank *U519* on 10th February, but the four other U-boats sunk during the month were destroyed near convoys, which reinforced Admiral King's conviction that U-boats were best dealt with there.

The Allies lost sixty-three ships (359,328 tons) to U-boat attack in February, an alarming advance on the thirty-seven ships and 203,128 tons of January. New boats more than made up for U-boat losses, and Dönitz began March with confidence.

Convoy SC121 was detected by the German Radio Intercept Service not long after it had left New York. Thirteen boats of the Neuland group gathered south of Greenland's Cape Farewell along the 59th parallel. By the time they made their heaviest attack on the night of 9th/10th March the convoy had been scattered and battered by a succession of south-westerly and westerly gales, with heavy, savage seas. In the eight escorts, four had radar breakdowns, three lost the use of their asdics, HF/DF sets were knocked out, R/T became unreliable. Twelve ships were sunk.

Three more convoys followed at rapid intervals. They were all routed three hundred and fifty miles further to the south than SC121, but Radio Intercept picked them all up.

The first of these, HX228, left St John's, Newfoundland, at 6.45 p.m. on 5th March. Escort Group B3, comprising the British destroyers *Harvester* (Commander 'Harry' Tait, RN, as SOE) and *Escapade*, the corvette HMS *Narcissus*, the Free Polish destroyers *Garland* and *Burza*, and the Free French sloops *Aconit*, *Renoncule* and *Roselys*, met the merchantmen at ten o'clock on the morning of the 6th, and at 1.45 p.m. USS *Bogue*, with her destroyers *Belknap* and *George E. Badger*, of Task Unit 24.4, came up with the convoy. Captain Giles E. Short, USN, steered his ship into a central station among Commodore J.O. Dunn's sixty-one ships, the first carrier to join a North Atlantic convoy since *Audacity*, nearly eighteen months before.

Bogue was a big advance on the little British carrier. She had a hangar and a catapult, and carried many more aircraft – the twelve Wildcat F4F4 fighters and nine Grumman Avenger TBF1s of Composite Squadron (VC) 9, under Lieutenant-Commander William McClure Drane, USN.

The Avenger TBF was a modern torpedo/bomber, an all-metal plane with a strong undercarriage for deck landing, its three-man crew of pilot, radioman and tail gunner enclosed in a 'glasshouse', a top speed of over 220 knots, and a maximum range of 1,390 miles. In comparison, the Royal Navy's Fairey Swordfish was antiquated. This fixed-undercart wood and fabric biplane with open cockpits was obsolete in 1939. With no warload a 'Stringbag might just make 125 knots, 770 miles maximum range. But the old machine had advantages in the anti-submarine war. It was easier to sight a U-boat from its open cockpits, and the models with which the current British, American-built, 'escort carriers' were equipped had better versions of the new ASV (Air to Surface Vessel) radar than the Avenger, and some of HMS *Archer*'s Mark IIs were fitted to operate the new rocket projectiles, as yet untried against U-boats.

Bogue flew dawn-to-dusk Avenger patrols on 7th March, then Allied Intelligence in their turn located the concentration of U-boats ahead, and the convoy was ordered to make a forty-five degree turn to starboard. Heavy weather prevented flying on the 8th and 9th, and the escorts' HF/DF, with which *Bogue* was not fitted, reported U-boats signalling to the north.

At 2.32 p.m. on the 10th Kapitänleutnant Hunger in *U336*, southernmost boat of the Neuland line, sighted the convoy and radioed Lorient, then closed to shadow the enemy. Huff-Duff in the convoy escort heard his signal, and Commander Tait asked *Bogue* to send planes to investigate. Short catapulted a TBF.

Lieutenant (jg) Alex C. McAuslan was flying in mist and rain at 12.26 when he sighted Hunger's boat on the surface to starboard. He called up the ship but could not get through, then opened the throttle and said over the talkback, 'This is it. Man the turret. Check your switches and break out the camera.' Gunner Boyd wriggled up through the manhole and crouched inside the ball turret, hand on the big .5 calibre machine-gun. Radioman Newman collected his camera and climbed down into the after belly position, next to his .30 gun, ready to take pictures.

Hunger was not expecting aircraft, and the big Avenger in its duck-egg blue camouflage was in its final power glide before he sighted it. *U336* crash-dived, and her periscope standards were still above water when McAuslan pressed his depth-charge release button. Three times he tried, but the release malfunctioned, though he did alert the destroyer *Escapade*, which steamed across and kept the shadower down. In the afternoon the turret gunner of another TBF returning from patrol sighted the swirl of a diving U-boat two miles from the carrier. Again the DC malfunctioned, and only one charge, with a faulty fuse, was dropped. The *Bogue* group then left the convoy to return to base, to assess this preliminary sortie and amend mistakes.

Oberleutnant Langfeldt, whose *U444* had taken over as shadower from the nobbled Hunger, saw them go, but had no means of knowing whether it was for good. Oberleutnant Trojer's *U221* waited four hours until dusk before she attacked the convoy from ahead submerged, sinking the two freighters *Turinca* and *Andrew F. Luckenbach*.

Seeing the fiery columns in the darkness, Harry Tait put *Harvester*'s helm hard over and steamed back from ahead to where the stricken ships now were. Langfeldt had just put a torpedo into a third ship, the SS *Lawton B. Evans*, when Tait sighted him and went to the attack. *U444* crash-dived. *Harvester* started depth-charging.

Nothing came of this attack, but HF/DF was reporting more U-boats to the north-east, and four hours later Hunger's *U336*, which had got into the act again, *U86* and *U406* made attacks from both flanks of the convoy.

Tait saw the American *William G. Gorgas* and the British freighter *Jamaica Producer* hit. Then his lookouts reported a submarine on the convoy's starboard beam.

This was Langfeldt's *U444* again. She crash-dived once more, and again *Harvester* saturated the sea around her with DCs.

This time she could not escape. Leaking badly, *U444* surfaced, Langfeldt led her gun's crew out of the hatch, and opened fire on the destroyer, with bridge-mounted machine-guns as well as the 8.8cm deck gun. *Harvester* replied with her forward 4-inch guns, and both antagonists scored hits as the destroyer raced down on the U-boat. Langfeldt could not dive, and his manoeuverability was restricted by damage, and *Harvester* smashed into her and rode over the canting hull of the U-boat, badly damaging her own bows and side plating.

For some twenty minutes *U444* remained jammed under the destroyer's propeller shaft. In this embarrassing position neither vessel could attack the other. *Harvester* could obviously not fire any depth charges, nor the U-boat her torpedoes, but an underwater explosion put the destroyer's port engine out of action, and when the U-boat finally slid clear and made off slowly into the darkness her attacker lay disabled, unable to make way.

Racing to the scene was FFS *Aconit*, commanded by the aggressive Lieutenant de Vaisseau Levasseur. The voice of the radar operator rasped in the voicepipe . . .

'*Sousmarin! Sousmarin! A cinque cent metres!*'

Away to port Levasseur saw the U-boat, limping along. He put the helm hard-a-port and bore down on her, his biggest searchlight holding her in its beam, his oerlikons saturating her with 20mm

(*Left*) Commander Richard H. O'Kane, USN. (*Right*) Rescued US aviators aboard USS *Tang* 1944 after a carrier raid on Truk. (Lieutenant-Commander O'Kane centre).

USS *Tang*.

HM Submarine *Trenchant* which sank the Japanese cruiser *Ashigara*, 8th June 1945.

USS *Cavalla*.

shells. The bows of *Aconit* loomed over the almost helpless *U444*, and Levasseur had a split second to see men on her conning tower and casing before there was a grinding crash and shock as the sloop tore open the sub's pressure hull. As she sank, the Frenchman dropped five DCs, which finished her off.

Feeble cries of '*Hilfe! Hilfe!*' came from the water. *Aconit* picked up just four survivors from *U444*, and steamed back towards the convoy.

In the blackness towards the end of the middle watch *U757* made an attack on the convoy from the starboard flank. One of her torpedoes hit the ammunition ship *Brant County* and set her on fire. She burned brightly for some minutes, illuminating the whole convoy, then there rose a huge sheet of flame where she had been and a shattering explosion in which *U757* was damaged. The U-boat then came upon the disabled *William G. Gorgas* and finished her off with a torpedo.

At dawn the two Coastal Command Liberators *B120* and *R86* gave the convoy air cover once more. *B120* made four U-boat sightings during the forenoon, *R86* saw a surfaced submarine at noon, and all were attacked.

Meanwhile the crippled *Harvester* had managed to get under way and was struggling along at eleven knots. Then her only effective propeller shaft gave way and she drifted helplessly, awaiting a tow. She was still in this state when Eckhardt's *U432* came upon her in the afternoon and put two torpedoes into her. *Harvester* broke in two and sank rapidly with heavy loss, including Commander Tait.

Levasseur saw the black column of smoke which was her funeral pyre and altered course towards it. Three miles from the spot his asdic got a loud ping.

Aconit dropped twenty DCs in two attacks and tossed twenty-three missiles from her new Hedgehog anti-submarine launcher at the target. *U432* surfaced on her starboard quarter and Levasseur engaged her at once with his 2-pounder, then oerlikons, then his 4-inch. At about a mile and a half, closing, the latter scored four hits, one in the engine room, three on or near the conning tower.

Levasseur rang down Stop Engines, then Half Astern, intending to put his bows athwart the U-boat and send over a boarding party, but while the ship was still moving forward she rammed the U-boat, which sank immediately. *Aconit*, her bows only slightly damaged, but her underwater asdic dome smashed, stopped and hauled aboard twenty survivors, including *U432*'s First Lieutenant but not Eckhardt, who had been killed by the sloop's opening fire.

By this time Convoys HX229, from Halifax, and SC122, from New York, were at sea, eastbound. Some forty U-boats, including eleven of the Neuland group, were lying across their path, most of them still well stored with fuel and torpedoes. Radio Intercept had as usual tracked them from signals, and Lorient reorganised the boats into three new groups, Stürmer, Dranger and Raubgraf, Stürmer to attack HX229, the other two allocated to SC122.

HX229's ocean escort consisted of two destroyers, two corvettes and one destroyer for part of the time. SC122 had the destroyer *Havelock*, one frigate and five corvettes, with one US Navy destroyer part-time. Between 16th and 19th March HX229 lost thirteen ships (93,000 tons) out of forty, SC122 lost eight ships (47,000 tons) out of fifty-one. But for HF/DF, radar and good air escort at each end of the voyages, the attrition would have been far greater.

On 20th March *Bogue* and her destroyers joined Escort Group B2, commanded by Commander Donald MacIntyre, who had captured Otto Kretschmer two years before, to support Convoy SC123. The passage was beset by every kind of bad weather, with snow, ice and fierce gales. High winds made continuous air patrols impossible, and *Bogue* had to leave the convoy prematurely as the rough weather made it too dangerous to try to refuel her thirsty old DDs from her own capacious fuel tanks The equally ancient V-and-W destroyer HMS *Vanessa* crashed though heavy seas to depth-charge and seriously damage a U-boat which had been overheard by Huff-Duff urgently signalling a sighting report in German naval Enigma code.

The new German radar detector receiver being fitted in U-boats was helping to blunt RAF Coastal Command's offensive in the Bay of Biscay, but in March the new, undetectable 10-centimetre ASV radar and the 80,000,000 candle-power Leigh Light came into use. An aircraft would make a stealthy approach using its 10cm ASV to within a few hundred yards, switch on the Leigh Light and attack before the submarine knew it was there.

On 22nd March a Wellington sank *U665* at night in the Bay, but the same day *U338* shot down an Australian-manned Halifax, and the U-boats began 'fight back' tactics, ordered by Dönitz, armed with the new 37mm twin, 20mm twin and quadruple quick-firing cannon and heavy machine-guns, mounted on special platforms, called *Wintergärten*, or 'bandstands', on the conning tower aft of the bridge.

For the first twenty days of March ships sunk by U-boats rose to the record horrifying total of 627,377 tons, or 108 ships lost, two-thirds of them in convoy, which, the Royal Navy Chief of Staff recorded, 'It appeared possible that we should not be able to continue as an effective system of defence.'

In April the first British-manned escort carrier to support a convoy in North-Atlantic for well over a year covered Convoy ONS4 to Argentia, Canada. She was HMS *Biter*, an American conversion with a small Swordfish squadron aboard. While ONS4 was at sea, USS *Bogue* and her four old four-piper destroyers left Argentia to escort eastbound HX235.

Escort commander Donald MacIntyre was expecting *Biter* to travel inside the convoy and provide close air patrols to spot and attack the U-boat patrol line which a spate of signals intercepted by his HF/DF told him lay dead ahead. But Captain Conolly Abel Smith operated the carrier and her escorts as a separate hunting group, covering the convoy only at a distance. In the vile weather, one of *Biter*'s Stringbag crews, flying through snowstorms in their open cockpits, sighted a surfaced U-boat, which dived before the old plane could reach her. MacIntyre was not informed, but sank *U191* without help from the carrier. Grudgingly Abel Smith flew a few patrols round the convoy, and on Easter Day one of these sighted *U203* and guided the destroyer *Pathfinder* to a kill.

Steaming further south in better weather, *Bogue* was able to top up her destroyer screen. Then the weather worsened and the small carrier, always susceptible to heavy seas, began to pitch and roll badly. In temporarily clear skies Lieutenant Roger Santee and his TBF crew sighted a surfaced U-boat fifty miles from the convoy and dived to attack, but his four DCs ricocheted off the water and there was no evidence of damage to the submarine, though hunting escorts found no trace of it.

In the Bay of Biscay all U-boats were forced to submerge at night. The Admiralty asked for 190 Lancasters to be switched from the strategic bombing of Germany to the Bay. Air Chief Marshal Slessor, the new C-in-C of Coastal Command, torn by conflicting loyalties, refused to endorse their request, and asked for six squadrons of Liberators from the USA. The US Navy wanted these aircraft for convoy defence, and refused. However, in the first week of May Sunderlands and Halifaxes sank four U-boats in the Bay. There was another kill on the 15th, and on the 24th a Sunderland badly damaged the special 'flak boat' *U441*, which was armed with rocket projectiles, 37mm cannon and machine-guns, but was itself shot down.

On 21st May *Bogue*, with five destroyers, was at sea five hundred miles south-east of Cape Farewell escorting the thirty-eight ships of westbound ON184. Lieutenant-Commander Drane and his aircrews were frustrated by their continuing lack of U-boat kills. While the carrier had been in Belfast having HF/DF fitted they had all gone on a course at the Coastal Command Anti-Submarine School at Ballykelly. Waiting for them over the horizon were the U-boats of the Donau-Mosel group.

Drane was flying on the last patrol of the day when he sighted the wake of a U-boat about sixty miles off the convoy's starboard beam, reported to *Bogue*, then pushed over and dived on the enemy. The U-boat dived too, and his four DCs had no noticeable effect.

Next day dawned clear, with visibility fifteen miles, and some useful cloud at 1,500 feet from which to stalk U-boats. Just after half-past six Lieutenant (jg) Roger Kuhn's Avenger crew saw *U468* surfacing three miles off their port bow. When he attacked, the submarine opened fire with 20mm cannon. Kuhn's own fire hit the enemy gun crews and stopped the barrage, and he dropped four DCs. The U-boat circled, trailing a long bluish oil streak, and there was a flash of intense white light from her stern. After about an hour she submerged stern-first, though no wreckage came up.

Lieutenant Dick Rogers, coming up in Wildcat to assist, also saw the oil. Kapitänleutnant Rudolph Bahr stayed down until he was sure the fighter had gone, then surfaced and continued his pursuit of the convoy. He and the leading ships of ON184's screen closed each other head-on and were about eighteen miles apart when young TBF pilot Stewart Doty, flying through a rain squall, saw *U305*'s long white wake about nine miles off his starboard bow.

Doty climbed and overtook the U-boat in cloud. When he emerged he saw the submarine dead ahead of him, travelling fast towards the convoy, which was about an hour's steaming away.

Bahr had got complacent, and the lookouts did not see the bulky Avenger until it was coming in on its attack run, then one man ran to a 20mm gun on the conning tower and opened fire. Doty hung on, lowered his wheels to steady the plane down, and let go four charges. Radioman Pollack, clicking his camera shutter at the ventral window, saw them go off all in one spot close on the U-boat's port quarter. Her hull jerked to port, then to starboard, and she stopped, then slowly settled below the waves. Out of the ocean rose a huge blue oil bubble fifty feet across. About two minutes later part of the sub's conning tower and twenty feet of her bow came up at an angle, then slowly slid below again, and belched up a second huge blotch of oil. Pollack shouted, 'We got her!'

U305, her pressure hull damaged, was still functioning. Bahr took her down as deep as he dared, made temporary repairs, and surfaced again to overtake the convoy.

He got to within twenty-six miles of ON184's starboard quarter when Lieutenant R.L. Stearns' TBF dived on him, cowl gun firing. Bahr replied with 20mm shells. Stearns dropped a DC salvo, his turret gunner raked the U-boat with his 50-calibre.

Bahr submerged again, but now at last he had had enough. Eluding the hunting US destroyer *Osmond Ingram*, he made more emergency repairs, and just managed to get back to Brest.

To the south of *Bogue*'s convoy steamed eastbound HX239, with escort carrier HMS *Archer*, convered from an American freighter, in its escort.

The wind over the deck had sunk to only seventeen knots when Swordfish F took off, and she could only lift two depth charges. Twenty miles off the convoy's port beam her crew sighted *U468*, Roger Kuhn's former opponent, which had given up ON184 and was now looking for the more southerly convoy, reported by Radio Intercept. The Stringbag stalked the U-boat from cloud cover, then dived on the enemy from two miles ahead. The U-boat contemptuously made no attempt to dive, but went into a zigzag. *Archer* radioed the plane that another Swordfish and a Martlet were on their way, but the sub dived before they could get there. The first Swordfish dropped her two DCs, but there was no sign of a hit.

At 5.23 p.m. *Bogue*'s brand-new HF/DF picked up a U-boat transmitting Enigma twenty-three miles off ON184's port quarter. Half an hour later Lieutenant(jg) William F. Chamberlain's *TBF* was captapulted off to head down the bearing supplied by Huff-Duff. Seven minutes after launching he sighted the wake of *U569*.

Diving out of cloud cover he ran up the sub's stern and let her have his four DCs from a hundred feet. Don Clark opened up with his .5 from the turret and saw the four charges make a perfect straddle, two each side of the submarine.

In five minutes the Avenger of Howard Roberts was on the scene, guided in by Chamberlain's smoke markers. Chamberlain, on his way back to *Bogue*, heard him shout.

'The sub's coming up!'

Chamberlain turned round and raced back, in time to see Roberts' plane diving on the U-boat as she steered west into the sun.

His DCs straddled her stern. *U569* lifted, fell back, heaved upwards again, turned over on her side, sank once more, and rose again on an even keel. Men started jumping out of the conning tower and galley hatches, and both *TBF* turret gunners opened up to drive them back into the boat, so that they would not be keen to open their valve. No more heads appeared in the hatches. Then the gunners ran out of ammunition. While they were fitting fresh cans about thirty Germans came out on deck, some of them waving a white table cloth. The gunners opened fire again, and some men fell back down the hatches, other jumped overboard.

As the two planes circled, the Canadian destroyer *St Laurent* steamed up, stopped and lowered a boat with a boarding party. Seeing them, the U-boat's Engineer Officer, who was among some dozen men still aboard her, slid below and opened all the flood valves, and in a few minutes the submarine sank. Twenty-four men were rescued from the rough sea.

The Americans had scored. It was also the first time a U-boat had been sunk directly by aircraft from an escort carrier.

The boats of the Donau-Mosel group were scattered by now, but *U218* sighted HX239 seven hundred miles north of Flores in the Azores. The nearest boat was Karl Schroeter's *U752*, and Lorient orderd him to attack the eastbound convoy.

Schroeter had sunk nothing on either of two previous patrols, and he made all speed towards HX239 through the night. At dawn he signalled Lorient. Commander Evans' Huff-Duff operator in the destroyer *Keppel* picked it up. She headed at full speed down the bearing, and HMS *Archer* flew off two Swordfish. They sighted nothing, but at 8 p.m. two escorts got good bearings from their HF/DFS, and *Archer* sent Swordfish F to investigate.

The Stringbag sighted the U-boat ahead. She dived but kept her periscope up to watch a circling Liberator, and the Swordfish dropped four DCs in a straddle.

Swordfish G and Martlet B relieved F and sighted a U-boat ahead of them on a reciprocal course, creaming along at a good twelve knots. From its unusual off-white colour and large conning tower it looked like one of the big 'milch cow' supply boats which Dönitz had been sending out since the spring of 1942 to keep his boats longer on patrol. Both aircraft attacked at once out of a cloudless sky. The U-boat turned hard-a-starboard and dived. The Martlet's bullets rattled on her stern gratings. The Swordfish dropped four DCs ahead of the swirl.

Twenty minutes after this attack Swordfish B, piloted by Sub-Lieutenant Harry Horrocks, with Sub-Lieutenant Noel Balkwill as his observer, took off from *Archer*. They were flying off the convoy's port quarter awaiting orders when they sighted Schroeter's *U752* ten miles off, heading at a fast fifteen knots for the convoy.

Horrocks turned into cloud, and Balkwill passed him on a course which he estimated would bring them to a good position for a surprise attack. A Stringbag needed all the advantages it could get. The Swordfish flew on in the grey smelly murk for four and a half minutes. Then Balkwill shouted through the speaking tube, 'Break cloud now. She should be dead ahead, range about a mile.'

The Swordfish's drooping fixed undercarriage emerged out of the cumulus at 1,500 feet. There was the U-boat, very fine on their port bow. Horrocks dived, keeping the sub in his old ring-and-bead gunsight.

At eight hundred yards he flipped the firing switch, and the first pair of RPs shot from their racks under the lower wing. Schroeter saw them splash a hundred and fifty yards short. He pressed the klaxon button, and the boat tilted in a crash-dive.

At four hundred yards Horrocks triggered the second pair. They were nearer but still short, and the conning tower was nearly down. The Stringbag jerked as the third pair leaped forward. Ten feet short.

There were some twenty feet of the U-boat's tilting hull showing, rudder in the air, when the last pair of RPs smashed into the hull on the waterline.

U752 slid on down for a few minutes, then rose slowly and circled to port, trailing an oil slick. Men leapt from the conning tower hatch and manned the after 20mm gun, but Horrocks kept the frail Stringbag out of range. The pilot of Martlet B, on his way back to the carrier, heard Horrocks' request for back-up, and in a minute was diving on the U-boat. His last burst of 50-calibre fire at the conning tower killed Schroeter. The destroyer HMS *Escapade* arrived a few minutes after *U752* sank, and picked up thirteen survivors. Ten more were rescued some hours afterwards by *U91*.

These successes in May were symbolic of the dramatic change in fortunes on the U-boat front. During the month the Allies lost 45 ships to U-boat attack, a total of 265,000 tons, about a third of the March total. The really bad news at U-boat Headquarters was the loss of forty-one boats and over two thousand highly trained men. It dealt a blow to the morale of the U-boat service from which it never really recovered, and foreshadowed a final victory for superior Allied technology at sea.

On 24th May Grossadmiral Dönitz signalled to all his boats, 'The situation in the North Atlantic now forces a temporary shift of operations to areas less endangered by aircraft.' The 'areas less endangered . . . ' were the Central and South Atlantic and the Indian Ocean.

The Central Atlantic was the US Navy's responsibility. Here in the next twelve months US Navy escort carriers, destroyers and the new, fast destroyer-escorts savaged the U-boats, using radar, sonar, HF/DF and anti-U-boat missiles old and new. Most effective was the new acoustic homing torpedo known for security reasons as a Mark 24 mine, but more familiarly as Fido or Wandering Annie. Torpex-filled Fido homed on the noise made by a vessel's screws as they churned the water, using a hydrophone in its head, shielded so that it did not chase its own tail. With a range of 1,500 yards, Fido would continue to search for fifteen minutes beyond that distance and many U-boats were fatally bitten by it.

Added to the great assistance given to the U-boat hunters by Huff-Duff and radar was the decrypting service provided by Admiralty cypher specialists and especially the British Government Code And Cypher School at Bletchley Park. Capture of the Enigma coding-decoding machine from Lemp's *U110* and the cyphers from the *Munchen* had helped Bletchley break the naval M-Home Waters Hydra code used for communications between U-boats and base, and it began to be possible to re-route convoys clear of U-boat packs. In February 1942 the Germans for added security switched to the new Triton code. In December 1942 Bletchley broke Triton, and the results gradually showed in the location of U-boat packs. A delay in British decrypting and an upsurge in the efforts of the German Radio Intercept Service in March 1943 partly accounted for the U-boat victories in that month, but the ultimate decoding machine known as the Bomb at Bletchley was back on song in April.

A wolf pack, unlike an American carrier task unit, was not an independent roving force. From the moment she put to sea a constant flow of radio signals controlled every movement a U-boat made. Shortly after she had left port she was told where to go and when she was expected to arrive there, and had to send passage reports after she had cleared the Bay. Patrol lines were always formed strictly on the hour, each U-boat commander who was to be in the line was addressed by name and told his positions in the line at stated intervals defined down to minutes by means of grid squares. Detailed descriptions of situations were transmitted by each U-boat on frequencies heard equally well on the US side of the Atlantic.

If a U-boat did not regularly report its position or failed to transmit for several days, the commander received a special request to report. Losses were judged, by both sides, on this basis. A U-boat could only start its return home when ordered to do so or after a request for special consideration had been granted. Only if a boat was badly damaged was its CO permitted to make his own decision to return home, and then only after he had reported his intention. On his way back he had to transmit his expected time of arrival off escort points, and his request for radio beacons.

A U-boat's fuel state had to be included in every transmission. A refuelling station at sea was always announced well ahead of time, together with the number of particular boats to be serviced, how much fuel they should have on arrival, and how much they were to be given, all down to the last cubic metre. When a refuelling session was over, the milch cow was required to report which boats she had refuelled and in what order, exactly how many cubic metres each had received, and an inventory given of her own remaining supplies.

The U-boat channels carried many technical details to or from boats wanting advice on modifications or repair, and were choked with far more general orders, estimates, situation reports, information on individual crewmen's ailments and proxy marriages than was necessary for the day-to-day operaton of a boat, thanks to the German love of book-keeping.

All this information, of priceless value to the Allies, was entrusted to the radio, and much of it was picked up and turned into Ultra intelligence.

One of the first instances of the US Navy's offensive use of Ultra was the very successful attack by *Bogue* on the Trutz Group's patrol line in early June 1943. After that there were frequent attacks based on decoded Enigma, with special attention given to refuelling rendezvous in Central Atlantic. By August 1943 Dönitz had lost nine of his original twelve supply boats, the heavily armed floating emporia which carried, in addition to fuel and torpedoes, such provisions as fresh strawberries, pears, asparagus and many other fresh vegetables, roast chickens, rabbits and brandy. These stores were transferred by means of a float in seas of up to Force 5, otherwise by strong manila from bridge to bridge, with the two submarines lying about thirty to forty yards apart.

Of the 489 U-boats sunk by Allied action at sea beginning January 1943, US Navy forces sank approximately sixty-three with the direct use of Ultra information, and some thirty more with its indirect aid. Details of many U-boat weapons, deception devices and operating techniques were gained. Among these was the secret of Aphrodite, the radar decoy device. This consisted of a small hydrogen-filled balloon about 65cm in diameter to which were attached silver wires tuned to radar frequencies. When a U-boat using her radar receivers discovered that she was being hunted by Allied radar, she could head into wind and release the balloon, which was carried to leeward of her, drawing the hunting craft away.

In June 1943 British Intelligence realised that the Germans had, in their turn, cracked the British naval codes used to brief convoys on the location of U-boats. The codes were changed at once, and the new ones were never broken.

The Allied Bay offensive was now at its height. On 28th July began the six days of what came to be known as the 'Big Bay Slaughter'. On 28th July *U404* was sunk by an Anglo-US Liberator team. Next day a Leigh Light Wellington sank *U614*.

On the calm clear morning of the 30th a US Liberator spotted *U461*, *U462* and *U504* together, outward bound from Bordeaux. The first two were milch cows en route for the Cape Verde Islands to refuel U-boats making for the Indian Ocean. A Sunderland and a Catalina were called in. The Catalina went off looking for Captain Walker's hunter-killer 2nd Support Group, the Sunderland stalked the U-boat trio and homed in a US Liberator and a British Halifax. The U-boats turned in circles and put up a thick barrage. A second Halifax attacked with a Fido homing-torpedo, which holed *U462*'s pressure hull. An Australian Sunderland came in at periscope height and straddled *U461*. The big milch cow sank, leaving thirty survivors; a Halifax finished off *U462*.

Meanwhile Walker was on the scene, his flagship HMS *Kite* flying the 'General Chase' signal, hardly ever seen since Nelson's day. *Kite* sank the 740-tonner *U504* after a 'creeping attack' lasting two hours. *U454* and *U383* were sunk by Sunderlands on 1st August, *U106* by three Sunderlands, and *U706* by Captain Hamilton's USAAF Liberator on 2nd August. Nine U-boats had been sunk in one week, sixteen in a month. Dönitz compounded his defeat by withdrawing the excellent Metox radar receiver, which he mistakenly thought was betraying its presence to Allied search receivers.

In September U-boats began using acoustic torpedoes, and had some renewed success at first. Then the Allies introduced the 'foxer', a noise-making device trailed in the wake of a ship to explode the homer prematurely, and stepped up their hunting.

In October USS *Card*, *Core* and the new carrier USS *Block Island* sank six U-boats. *Card*'s planes got four of these, including the 1,600-ton supply boat *U460*, two of them being sunk on successive days. Other aircraft sank ten more, surface escorts a further ten. Also in October Terceira in the Azores was established, by agreement with Portugal, as an Allied air base.

On 5th November Stringbags from another escort carrier, HMS *Tracker*, guided Captain Walker's 2nd Support Group to the attack which sank *U226*. On the 29th Avengers from *Bogue* sank *U86*.

The successes of the US Navy escort carriers continued through 1944. The days of the big wolf pack had gone, and targets dwindled as the year dragged on, but five US CVEs, as these carriers were now designated, including the veterans *Bogue* and *Block Island* and the newcomers *Guadalcanal*, *Croatan* and *Wake Island*, operating at different times, helped to sink seventeen U-boats, either with their aircraft alone or in collaboration with surface escorts.

On 9th April *Guadalcanal*'s hunter-killer group damaged Werner Henke's *U515* and he blew tanks and surfaced in the middle of the hunters. Henke was a hard-nosed veteran, who had sunk 150,000 tons of Allied shipping, including the liner *Ceramic*, but untypically he put up no resistance at all when he came up. The resourceful and aggressive Captain Dan Gallery of *Guadalcanal*, unofficially known as the *Can Do*, saw this as probably indicative of a general weakening of U-boat morale, and thought that but for the fierce barrage which his group had laid on the surfacing *U515* he might well have captured her.

When the *Can Do* left Norfolk, Virginia, on 15th May for their next cruise Gallery had made plans to take over the next U-boat which he might have the luck to flush. 'We shall cease firing anything heavy which might damage her hull,' he told his commander. 'We'll just keep them away from their guns with small-calibre fire and drive them into the water so that we can get a boarding party over . . Then we'll pass a towline and bring her back to the USA.'

On 29th May he heard that *Block Island* had been sunk by a U-boat. Then he got a report of a U-boat homeward-bound between the Cape Verde Islands and the coast of West Africa, and sent his planes searching for her as the carrier herself headed for Casablanca in order to refuel.

Oberleutnant Harald Lange was inexperienced. First he mistook aircraft being ferried from Brazil to West Africa for the carrier task unit which was actually at that time some distance away. His movements to avoid this non-existent force helped to direct him towards Gallery's hunters. Lange grew more and more rattled. Every time he tried to surface and recharge batteries his Naxos radar detector receiver screeched its warning of aircraft, and he had to dive again. This did nothing for the morale of his crew, which had been low enough when he took over, as a series of unhappy incidents, including the suicide of one commander while on patrol, had given *U505* the reputation of an 'unlucky' boat.

Dan Gallery pushed his own luck. With *Guadalcanal*'s fuel state critically low he persisted in the hunt, and on the beautiful, clear morning of Sunday 4th June the destroyer escort *Chatelain* got a ping. After twenty Hedgehogs and twelve DCs Lange brought *U505* up.

Following the plan, the American gunners forced the already demoralised German sailors off the submarine with oerlikon and machine-gun fire, some of which wounded Lange, and Lieutenant Albert L. David led a boarding party which managed to get below into the empty submarine. As the first American sailor leapt aboard her with a coiled rope, Gallery shouted over the Talk-Between-Ships microphone, 'Ride 'em, cowboy!'

Working against time, the boarding party closed off the most dangerous leaks, and discovered that none of the fourteen TNT charges intended for the destruction of the boat had been activated. *Guadalcanal* took the sub in tow and, thanks to a superb feat of inventive skill by *Can Do*'s Chief Engineer, Commander Earl Trosino, who had never been in a submarine before, brought *U505*, rechristened *Can Do Junior*, into Bermuda, whence she was taken to the USA. On the way, Gallery heard that the Allies had landed in Normandy.

The capture of *U505* had climaxed the victory of air power over the U-boats. Whenever a U-boat came up there seemed to be a plane overhead.

The Germans introduced the *Schnorkel* ('snort'), a short breathing tube through which a submerged U-boat could draw air to run the diesel engines and recharge the batteries without coming up, new Type XXI boats with an underwater speed of seventeen knots, the Balkon hydrophone, which could detect an enemy at fifty miles, and the new S-gear, a supersonic underwater detector which could do this even better. Experiments with the really revolutionary Walter-boats, which used hydrogen peroxide fuel, had been successful, and one hundred of these phenomena, which could maintain a speed of twenty-five knots underwater for twelve hours, were begun in May. Heavy Allied bombing raids, however, began to disrupt production schedules.

In mid-March 1945 two of a new small 200-ton U-boat Type XXIII, the modern equivalent of the old 'canoes', but with an underwater speed of thirteen knots, shoved off from Christiansand in Norway. Off the east coast of Scotland Barschkies' *U2321* sank a freighter. During the following month six more XXIIIs sailed on experimental cruises, and their captains found them fast and agile enough to elude asdic and DC attacks. But like the XXIs and the Walter-boats, they had come too late.

In April, the first, and, as it transpired, the last of the big XXIs, *U2511*, sailed under the veteran Korvettenkapitän Schnee, who had formed a plan for a surprise attack on Panama. On his way he was picked up by asdics, increased speed to sixteen knots, and easily got away. On 4th March *U2511* had her *Schnorkel* up when Dönitz's order to cease fire came in. During his return passage to Germany Schnee fell in with an Allied cruiser and four destroyers, manoeuvered into a perfect position for a shot at the cruiser, then with a great effort of will broke off and dived deep for home.

The second great war of the U-boats was over. Between 3rd September 1939 and 5th May 1945, they sank 2,603 Allied and neutral merchant ships, a total of just over thirteen and a half million tons, as well as 175 warships. Of 1,150 U-boats built before or during World War 2, 781 were sunk by enemy action, and 215 scuttled by the Kriegsmarine at the end of the war. After the surrender, 154 submarines were turned over to the Allies, though two, *U530* and *U977*, escaped to the Argentine.

CHAPTER TWELVE

Down The Throat

When Nagumo's planes hit Pearl Harbour on the morning of Sunday, 7th December 1941, the US Navy had 111 submarines in commission. Thirty-eight of these were old, S class boats built between 1918 and 1924. The US Navy regarded these as 'coastal' types, though some had a good operational radius on the surface. Most of them had four torpedo tubes, all forward, and twelve torpedoes. Four boats had five tubes, with one aft, and fourteen torpedoes. One 4-inch deck gun was fitted. They could make some fourteen or fifteen knots on the surface, ten to eleven knots underwater, and a crash-dive took all of a minute. Thereafter the Navy had gone for bigger boats, all named after fishes. The 2,000/2,506-ton *Barracuda*, *Bass* and *Bonita* of 1924-25 featured for the first time combined diesel/electric drive on the surface, and with this power could raise nearly twenty knots. *Argonaut*, completed in 1928, was a big minelayer with two 6-inch guns, *Narwhal* and *Nautilus* were large 2,730/3,960-ton submarine cruisers. After these came the development of the standard Fleet boats of some 1,450/2,200 tons, with six, eight or ten torpedo tubes, one 3-inch deck gun, a top speed of twenty knots on the surface, and a range of 12,000 miles. This type made up the majority of US Navy submarines in December 1941. Sixty-five more were being built under the 1940 Supplementary Programme, and after Pearl a further large number were laid down.

With two of the eight battleships caught at Pearl Harbour destroyed, three or more become mere hulks, and other ships, including three cruisers, badly damaged, the USA was temporarily the 'weaker nation' in the Pacific, and as such would be relying for some time to come on her submarines and carriers, particularly as her remaining capital ships would soon have to be divided between two oceans.

On 10th December 1941, most of the twenty-eight submarines, including six of the old S-boats, in Admiral Thomas C. Hart's Asiatic Fleet were already at sea, and were not caught on the 10th by heavy air raids on their Cavite, Manila, base.

The Japanese landed troops at Aparri in northern Luzon in the Philippines on the 10th. Lieutenant-Commander Freddy Warder took *Seawolf* into the bay, where the big 12,000-ton seaplane tender *Sanyo Maru* lay at anchor. Warder fired four torpedoes at her. He kept the periscope on the target, and thought he heard the thud of an exploding warhead.

In fact one torpedo did hit but failed to explode. This was just the beginning of a long series of torpedo failures which drove skilled and aggressive American submarine commanders to baffled frustration. US submarines for some time used both the old Mark 10 contact torpedoes and the newer Mark 14, which were meant to detonate either on contact with the ship's hull or by the 'influence' of the ship's magnetic field on a magnetic actuating pistol, or 'exploder', in the warhead. All the torpedoes fired at the *Sanyo Maru* were Mark 14s, forerunners of all the rogue 14s which went off prematurely, passed under the target, or simply thumped it without detonating. Both marks had suspect depth-keeping mechanisms, but the older fish was far more reliable. U-boats had had troubles with their torpedoes in the battle for Norway, but they were not as serious as the problems which plagued the Americans.

On 2nd January Manila surrendered. In mid-February Singapore fell. On the 27th a force of five cruisers and ten destroyers, the last Allied surface warships in the East Indies, was destroyed in the Java Sea. Java fell, and with it the Allied submarine base at Surabaya. New bases were set up at Fremantle in south-western Australia and Brisbane on the east coast.

The submarines and a few carriers now held the line which stretched across the Pacific from Australia through the Solomons and Midway Island to the far Aleutians. On 31st January a task force based on the carriers *Yorktown* and *Enterprise* began raids on Japanese island bases. The subs *Pollack* and *Plunger* sank ships off Tokyo Bay. New submarines joined ComSubPac in February. On 31st March Warder's *Seawolf* sank the cruiser *Naka* off Flying Fish Cove, Christmas Island, two hundred miles south of Java in the Indian Ocean.

Other boats patrolled the Gilbert and Marshall Islands in the Central Pacific, Truk and the Carolines, the Marianas and the inshore waters of Japan, the Coral Sea off north-eastern Australia, the East and South China Seas and the East Indies, in spite of continued troubles with Mark 14 torpedoes.

Anderson's *Thresher* sank a *maru* south of Tokyo, and on 16th April pulled out of range of Japanese DF and sent Admiral Halsey, steaming away to the eastward in *Shangri La* (USS *Hornet*), the message he was waiting for. It was a weather report, and it gave General Jimmy Doolittle's sixteen Army B25 bombers the green light to leave *Hornet*'s crowded flight deck for Tokyo.

On 7th May US and Japanese carrier forces clashed in the Coral Sea. The Japanese carrier *Shoho* was sunk, and the bigger *Shokaku* damaged. The submarine *Triton* chased her in vain.

Submarines lying off Japanese bases gave warning of another big build-up of shipping for a major operation, and American Intelligence calculated correctly that the objective was the capture of Midway. Naval forces were deployed accordingly between Midway and the Aleutians to the north.

Thirty-six submarines were involved. On an arc west of Midway itself were grouped eleven boats. On 3rd June a Navy Catalina flying boat sighted the fleet of transports, freighters, battleships, cruisers and destroyers. Later a strike force of four carriers, its objective to crush resistance by Midway forces and the US fleet, was discovered about 150 miles north-west of the island.

Early on the 4th planes from Midway attacked the Japanese strike force and hit several ships, including the carrier *Soryu*. While they were in action, planes from their targets were bombing and strafing Midway.

At 8 a.m. the carriers *Enterprise* and *Hornet* launched their planes against the Japanese carrier force, followed by the *Yorktown*'s squadrons. At 8.24 a.m. Brockman raised the periscope of the old sub cruiser *Nautilus*, and found himself in the midst of the enemy fleet. A battleship on his port beam sighted his periscope and fired her big guns. Brockman fired two torpedoes. One misfired, the other missed. Soon afterwards the American carrier aircraft attacked the Japanese force, and left the carriers *Kaga*, *Soryu* and *Akagi* burning fiercely and settling in the water. Japanese bombers inflicted fatal damage on the *York Town*.

At half-past ten *Nautilus* was on the surface with her aerial listening out when she intercepted a report of a damaged Japanese carrier. At noon she came up with the burning *Soryu*, which had taken three direct hits from the American carrier planes on top of the damage already received. About two o'clock *Nautilus* fired three torpedoes from a range of about 2,700 yards, and they all hit the target, which went down.

Destroyers forced the submarine down to three hundred feet, but

S *Stygian* took part in the
ck on the Japanese
ser *Ashigara*, 8th June
5.

Ashigara.

lass submarine with Mark
hariots and containers
unted aft.

The last U-boat, U505, captured by the US carrier *Guadalcanal's* task unit, 4th June 1944, in situ outside the Museum of Science and Industry, Chicago.

shortly after this Brockman sighted a second burning carrier. This was *Kaga*, which had been hit four times by dive-bombers, and was slowly sinking. He fired three torpedoes at her. Two of them missed. The third struck amidships with a dull clang. *Kaga* sank later, followed by *Akagi* and *Hiryu*, the remaining two carriers.

Midway was a crushing defeat for the Japanese, destroying two-thirds of their crack carrier force, and ending their period of expansion. From then on the Americans assumed the offensive, with increasing strength.

The submarines formed the front line, decimating Japan's merchant marine and making more gaps in her already depleted warship strength. Ferrall's *Seadragon* destroyed four big *marus* in the South China Sea, Stovell's *Gudgeon* torpedoed two big tankers and a freighter west of Truk.

On 7th August the Marines landed in the Solomons. On the 10th the seventeen year-old S44, with worn-out machinery and obsolete fire-control, but carrying dependable old Mark 10 torpedoes, sighted the four heavy ships of the Japanese Cruiser Division Six returning north in triumph from the Savo Island action in which they had sunk three American cruisers. The *Kako* was hit by four torpedoes and sank in five minutes.

While this act of retribution was being played, *Nautilus* and the old minelayer *Argonaut* were two days out of Pearl on a very different mission. Aboard the two boats were the two hundred men of 'Carlson's Raiders' under Lieutenant-Colonel Evans F. Carlson who were to raid Makin Island in the Gilberts to gain experience and data for the great amphibious assaults to be made upon the Japanese island strongholds of the Central and Western Pacific, stepping stones to Japan itself.

At 5 a.m. on 17th August the Raiders landed in rubber boats on the beach at Makin, with the submarines maintaining a shaky contact by portable radio. *Argonaut* had been partly chosen for the mission because she was fitted with two 6-inch guns, and at 7 a.m. Lieutenant-Commander Pierce was called upon to use them to knock out two enemy ships which Carlson had seen in the harbour on the far side of the island. *Argonaut* lobbed sixty-five shells over the trees into the lagoon at a range, blind, of 14,000 yards. To Pierce's amazement the Marines reported both ships sunk.

The landing party was supposed to return to the submarines at 7 p.m. At 8.30 seven of the nineteen boats came alongside, reporting that the others were finding it impossible to make any headway through the surf. At dawn Brockman took *Nautilus* in to within half a mile of the reef, and a few more boats were able to reach her. There an aircraft sighted and bombed her, and she and *Argonaut* were both forced to retire until dusk, when they both crept back to the beachhead, where Carlson directed them to pick up the remaining Marines near the entrance to the lagoon on the leeward side of the atoll. By midnight the last men were aboard, leaving twenty-one dead, planes and equipment wrecked.

In late August the aggressive Klakring brought the new *Guardfish* up amongst the sampans and trawlers off north-eastern Honshu. He sank two patrol boats with his 3-inch deck gun and four days later blew the bows off the *Seikai Maru* with a torpedo as she came out of Sendai. The Japanese did not expect American submarines off this foggy coast, where visibility was seldom more than three miles, where the Black Japan Current set north at two knots, and an inshore counter-current ran at the same pace. Klakring used his radar to feel out the land, his fathometer to sound the shallows, and exploited a talent for doing the unexpected.

He tried four fish on another *maru*. Three malfunctioned, the fourth hit. On 2nd September Klakring's torpedoes behaved themselves and broke the freighter *Teikyu Maru* in two. Two days later one slow-set torpedo fired from 5,000 yards put another *maru* aground. From ambush under the lee of a headland *Guardfish* then proceeded to sidle out and sink the 3,738-ton *Tenyu Maru*, the *Kaimei Maru*, 5,254 tons, and the 2,276-ton *Chita Maru*, making a bag for *Guardfish*'s first patrol of eight ships totalling 70,000 tons, one-tenth the total sunk by Pacific submarines in 1942, which considerably exceeded Japan's rate of construction.

The new year 1943 began badly for the Pacific submarines with the loss of an Old Contemptible. The crew of a US Army aircraft returning from a mission on 10th January saw a group of enemy destroyers under submarine attack south-west of New Britain. One ship was hit by a torpedo, and they saw explosions alongside two others. The undamaged destroyers then dropped depth-charges, and a submarine's bow surfaced at a steep angle. The destroyers circled round her, riddled her with shells, and she sank. This was the end for the old *Argonaut*.

One young submarine officer who felt the loss of *Argonaut* keenly was Lieutenant Richard O'Kane.* She was 'a monster, a continuous

---

* *Clear The Bridge!* by Rear-Admiral Richard O'Kane, USN (Retd).

challenge', but she was special to him because in her he had won his submariner's twin dolphins. She had a brush with Jap destroyers on the night of Pearl Harbour, but lack of air conditioning led to excess humidity and electrical fires. 'Nearly half her machinery became inoperative, but that did not stop *Argonaut* from carrying out her mission to defend Midway.' In fact she had been the only ship available to oppose an immediate assault.

It did not take O'Kane long to feel equal affection for his next ship, the brand-new *Wahoo*, 'half the displacement and with twice the power . . twice the *Argonaut*'s speed and manoeuverability.' She also had ten tubes to the old boat's four, and twice the number of tinfish to fill them.

The *Wahoo*'s first captain, Lieutenant-Commander W.G. Kennedy, left the ship after two patrols, and was succeeded by Lieutenant Commander Dudley W. Morton. He and Dick O'Kane were both ebullient, aggressive officers with a great appetite for action, and found a close affinity, which helped the younger man fit in with Morton's original ideas on submarine command. He divided the duties of the commanding officer in an attack, and made O'Kane 'co-approach officer'. In *Wahoo*, as in all US submarines of her type, all attack and fire control instruments were in the conning tower. From here Morton planned an attack, and O'Kane gave the orders which put it into practice.

Morton, known to some close friends and senior officers as 'Mush', was a big Kentuckian, a man full of life who was interested in many things, and liked to be involved in them all. With his big hands he produced some fine, delicate embroidery, and if a crewman wanted a badge sewn on, the Skipper usually did it for him. He believed in discipline, so specially necessary in a submarine, but he achieved it in a relaxed, friendly way.

Wahoo was ordered to patrol Palau, with a recce of Wewak harbour on the north coast of New Guinea en route, which normally meant a periscope peep from seaward of the reef at traffic and installations inside. But while they were on their way there *Wahoo* was signalled that the Japanese had landed at Wewak. Morton decided to go right in, though Wewak was not marked on his small-scale charts. He borrowed a school atlas from Motor Machinist's Mate Keeter, which included a general view of Wewak and its approaches, photographed and blew up this part, marked in what meagre information the Sailing Directions gave, which did not include depths of water or positions of reefs or shoals. In early dawn on 24th

January *Wahoo* entered Victoria Bay and headed down the twisting, reef-strewn nine-mile channel leading to Wewak harbour.

When she rounded the last corner into the harbour O'Kane reported masts, then a big destroyer. Both *Wahoo*'s periscopes, designed for a strictly Geneva Convention operation in broad daylight only, were narrow in diameter, with small lighting lenses, but Japanese Navy lookouts were exceptionally good, and the sub's thin broomstick was spotted.

O'Kane fired three torpedoes from 1,800 yards. The destroyer had increased speed and they all missed. Morton recalculated, O'Kane fired again, but the destroyer had seen their wakes, and these missed too.

The enemy was coming down on them fast. At 1,200 yards *Wahoo* fired one torpedo, which also missed.

The destroyer was looming in the lens. O'Kane awaited the order to crash-dive, but Morton said, 'Keep the scope up. Let's give her one down the throat.'

At eight hundred yards, with just time for the fish to arm, too close for the Jap to dodge, O'Kane fired. If he missed now, *Wahoo* was in big trouble.

But there was one big explosion, and when *Wahoo* surfaced, there was the *Harusame*, with her back broken. Morton's crew now began calling him 'Deadly Dudley'.

Two days later *Wahoo* was in her billet south of the western Carolines on the shipping lane from Palau down to the Bismarcks. With O'Kane at the periscope, Morton at the TDC (Torpedo Data Computer), thinking tactics, they dived for an attack on two freighters.

The first two torpedoes hit the lead ship, another slammed into the second ship's quarter. Four minutes later *Wahoo* came up and O'Kane saw a third ship on the same course as the others, with the second ship weaving unsteadily towards *Wahoo*. The torpedoes hit the newcomer, then the belligerent cripple almost hit *Wahoo*.

O'Kane put up the periscope and saw the third ship stopped, with soldiers swarming over the decks. *Wahoo*'s torpedo sank her, and Morton ordered O'Kane to shoot up the boats and barges which the third ship had lowered, crammed with troops. 'Every one of them could kill one of our boys on New Guinea,' Morton said. *Wahoo* sank all the biggest barges, ran over the smooth water towards the distant cripple, which had now been joined by a fourth ship, a tanker, hit the latter amidships and broke her back, then destroyed the freighter with her last two tinfish as she was within sight of a rescuing destroyer.

Just to the east of the *Wahoo* in the Bismarck Sea north of Rabaul Gilmore's *Growler* sank the *Chifuku Maru* on 16th January, hit a freighter on the 30th, and then got into a surface battle with the naval auxiliary *Hayasaki*, and rammed her. *Hayasaki* opened up with machine-guns on the submarine's bridge, killing one lookout and the second officer of the watch, wounding Gilmore and two other lookouts.

Gilmore ordered, 'Clear the bridge!' The quartermaster and officer of the watch scrambled down the hatch to the conning tower, dragging the two wounded lookouts with them. Gilmore called out, 'Take her down!' Badly hurt as he was, there was no time to save himself if the boat was to have any chance of escape. When *Growler* surfaced again, Gilmore and the two dead men had been washed away. *Growler* limped back slowly to Brisbane, eighteen feet of her bows twisted at right angles to port. Gilmore received a posthumous Congressional Medal of Honour.

1943 was a year of hard, stubborn fighting everywhere in the areas of Japanese conquest, in the Solomons, which by November were in American hands, in New Guinea and in the Gilberts. In all these areas American submarines were to be found, sinking the *marus* bringing men and supplies to the island strongpoints.

Everywhere their score rose steadily, and their losses were not small. Bole's *Amberjack* sank a big freighter off Vella Lavella in the Solomons, then was sunk on the Rabaul traffic lanes. *Grampus* went down before destroyers escaping from Tip Merrill's cruisers in The Slot. An oil slick and pieces of wood and cork were all that was left of MacKenzie's *Triton* after a convoy night attack near Rabaul in March. The 'pig boats' sank twenty-one *marus* that month, *Tunny* hitting a big freighter near Wake Islands, *Permit* scoring off Honshu.

To intercept any big Japanese force trying to attack the Gilberts invasion ships, Captain John Cromwell, Commander, Submarine Division 43, taking passage in *Sculpin*, was to co-ordinate seven submarines for a combined attack. The US submarine command had naturally studied the operations of German wolf packs in the Atlantic, and Admiral Lockwood, commanding US Pacific submarines, had sent two packs into the East China Sea in late 1943. Captain Momsen's *Cero*, *Shad* and *Grayback* sank three *marus* and damaged two more, and were followed into the Tung Hai by Freddy Warder, late of *Seawolf*, and his wolves. The score was modest, and both commanders believed they would have been more effective controlling from ashore, with the men on the spot having more flexibility, though the increased radio traffic, which had so often betrayed U-boats, was likely to be dangerous.

On 18th November *Sculpin*'s radar picked up a fast convoy. The destroyer *Yamaguma* depth-charged her and, leaking badly, she surfaced, to be met by a salvo of shells from *Yamagumo* which shattered the conning tower and killed Commander Connaway and his Exec. The senior surviving officer, Lieutenant Brown, decided to scuttle *Sculpin* rather than risk her capture, and ordered, 'Abandon ship.'

Captain Cromwell had been fully briefed on all plans for the invasion of the Gilberts and all the present and future locations of Central Pacific submarines. 'I know too much', he told Brown. 'If they torture me I might tell them everything.' With eleven other *Sculpin* men, dead or seriously wounded, he went down with the boat, and eventually received a posthumous Medal of Honour.

American leapfrogging strategy continued with the capture of the Marshall Islands in February 1944. As always, submarines were there, on offensive patrol and lifeguard duty. The *Guardfish* and the *Trigger* both sank a destroyer. The new *Tang*, commanded by Dick O'Kane, who had been saddened to bear of the loss of his old *Wahoo* on 11th October in the dangerous Sea of Japan, sank five *marus* in three days in support of a raid on Saipan in the Marianas, scheduled for invasion in June.

Under the command of Sam Dealey, the USS *Harder* was a great fighting submarine. Dealey, quiet, friendly, had an almost compulsive urge to attack. In *Harder*'s very first patrol in June 1943 she had knocked out seven ships, including a big transport and a tanker from one convoy, winning Dealey the first of his three Navy Crosses and his ship the amended name of *Hit 'em Harder*.

Hitherto Japanese *marus* has been the priority objectives for submarine torpedoes. On 13th April 1944 a signal from the C-in-C made destroyers the top targets, but *Harder* had anticipated him, and sunk the destroyer *Ikazuchi* with two torpedoes in four minutes.

Dick O'Kane was not pleased when his fighting *Tang* was put on lifeguard for the carriers' big raid on Truk, but relieved some of his frustration by pounding Japanese gun batteries ashore which were making it hot for *Tang* as she scraped along the reef en route for a downed pilot. Task Force 58 lost twenty-six planes and nineteen aviators, but the submarines rescued twenty-eight fliers from the sea.

Twenty-two of these grateful men enjoyed *Tang*'s hospitality. She was almost always within range of guns ashore, but carrier fighers helped out, and battleship floatplanes spotted for life rafts, alighted and towed them clear of fire until *Tang* could empty them.

At dusk on the evening of 6th June Dealey's *Harder*, on her fifth patrol, was threading the reefs of the Sibutu Passage between the Sulu Archipelago and North Borneo, heading north to rescue Australian Intelligence agents from a beach on the Borneo coast, when she picked up a convoy of three tankers heading south for the Borneo oil port of Tarakan. The escorting destroyer *Minazuki* sighted and attacked *Harder*, and at 1,100 yards Dealey fired a spread of four. Two of them hit, and the destroyer sank in minutes.

Next morning a floatplane forced *Harder* down and called in the destroyer *Hayanari*. Dealey let the range close to 650 yards, then fired three fish down the throat, Mush Morton style. The destroyer skipper threw his helm hard over, and the first torpedo hit him amidships, the second one aft.

The *Harder* carried on north through the Sibutu Passage, struggling against the current, with destroyers everywhere, picked up the six Australians, and returned down the Sibutu Passage for their main mission, a recce of the anchorage of Tawi-Tawi in the Sulu Archipelago, where Admiral Ozawa's fleet was moored. In the Passage she met and sank the destroyer *Tanikaze* and her consort.

As *Harder* approached Tawi-Tawi anchorage on the afternoon of 10th June she met the monster battlewagons *Yamato* and *Musashi* and their squadron of· cruisers and destroyers leaving to attack MacArthur's forces off New Guinea.

A destroyer sighted *Harder*'s periscope feather and raced in at thirty-five knots. Dealey fired three torpedoes down the throat, and she blew up.

He closed the anchorage that night to count the ships still there. When he reported his findings he was ordered to a quieter area, and Austin's *Redfin* brought in to replace *Harder*.

Austin was in his billet off Tawi-Tawi on 11th June when the *Asanagi Maru* approached with oil from Tarakan for Ozawa's ships, and *Redfin* sank her. Ozawa took council of his fears convinced that he was ringed round with submarines. His green aircrews, replacements for all the veterans lost in the bitter carrier air battles, badly needed flight training, but he kept the carriers swinging round the hook, growing barnacles. His fliers were still without practice when they were ordered to the Philippine Sea, which stretches east from the Philippines to the Marianas.

It was early morning when the carrier force sortied. Austin manoeuvered to get a shot, but the ships were steaming too fast and too far away for a a submarine at periscope depth. As soon as it was dusk, however, Austin surfaced and sent the signal which took Spruance's Fifth Fleet into the Philippine Sea, to put up a barrier of planes protecting the landings on the Marianas islands of Saipan, Guam and Tinian.

Admiral Lockwood had nine submarines in the western Philippine Sea watching for Ozawa. On 15th June the US Marines landed on Saipan. At 4.30 p.m. Risser in *Flying Fish* sighted the Japanese carriers coming out of San Bernardino Strait between Leyte and Mindanao. The new *Cavalla* sighted the Japanese task force on the evening of the 16th, came up and informed Spruance, who, at noon on the 18th, steered west through the Philippine Sea to meet the enemy. Having failed to contact Ozawa, he turned back at 8.30 p.m., worried for the safety of the force at the beachhead. But Ozawa's planes had located him, and at 7.30 a.m. on the 19th they launched their first strike.

Blanchard's submarine *Albacore* sighted a Japanese carrier at 7.46 a.m., followed by another carrier, cruisers and destroyers. He headed for the second carrier, which was steaming fast, and firing a spread of six torpedoes.

On the admiral's bridge of *Taiho*, the biggest carrier in the Japanese Fleet, Ozawa watched his planes taking off for their strike at the American carriers.

A Zero pilot left the deck, banked, saw a white torpedo wake heading for his ship, and dived on it, blowing kipper, plane and pilot to pieces.

But another torpedo hit. Normally one torpedo would not necessarily disable a big ship, but this one ruptured aviation gas tanks, full of the volatile Tarakan fuel. At 3.23 p.m. *Taiho* blew up and sank.

Three hours later *Cavalla* sighted the veteran carrier, *Shokaku*, damaged at Coral Sea and often hunted by US submarines, hit her with three torpedoes and sank her.

In late June O'Kane's *Tang*, Weiss' *Tinosa* and Reich's *Sealion* fell upon unprepared shipping in the Yellow Sea between Shantung Province and western Korea, regarded as a safe area by the Japanese. *Tang* sank ten ships, *Sealion* four, and *Tinosa* two.

It was a time for giant-killing, and the American submarines were in their heyday, decimating the already dwindling Japanese merchant marine, cutting out their carriers, cruisers and battlewagons, haunting the waters off Japan, called the 'Hit Parade', and the densely packed north end of the South China Sea, known as 'Convoy College.' Lee's *Croaker* and McMaster's *Hardhead* both got into the Hit Parade in August and sank the light cruisers *Nagara* and *Natori*, but south of Lingayen Gulf *Harder*, the destroyer killer, was sunk with all hands.

On 11th October McClintock's *Darter* and Claggett's *Dace* were patrolling the west coast of Borneo when they were ordered northeast to cover the Balabac Strait, off northern Borneo, and the Palawan Passage to watch for Admiral Kurita's battleship force breaking through towards the Philippines.

On 14th October *Dace* sank two of Kurita's tankers. At midnight on 20th October McClintock heard on the news that MacArthur's forces had landed on Leyte, and two days later the two submarines were coming down through the Palawan Passage when they picked up eleven of Kurita's heavy ships, which included the mighty *Musashi* and *Yamato*, three older battleships and seven heavy cruisers. *Darter* sank the cruiser *Atago* and damaged the cruiser *Takao*, *Dace* sank the cruiser *Maya* in four minutes.

Kurita's other ships steamed on towards the desperate and bloody engagements which formed the battle for Leyte Gulf and the American beachhead in the Philippines, to retreat with *Musashi* and three cruisers sunk by carrier planes, and the beachhead secure, though it was a close-run thing. Two three-boat wolf packs were in the Philippine Sea looking for cripples limping away from the battle. *Jallao* aimed seven torpedoes at the already damaged light cruiser *Tama* and sank her.

Dick O'Kane's *Tang* was on her fifth patrol. On 23rd October she surfaced at night in the middle of a convoy steaming through the Formosa Strait and in best Kretschmer fashion hit four of the five *marus*, sinking two and badly damaging the others. Twenty-four hours later she penetrated another convoy and sank two big tankers. An attacking destroyer received a Mush Morton down the throat and sank.

Then O'Kane had only two torpedoes left. He aimed one at a transport and saw it running straight towards her. Then he fired the second, their last. These Mark 18 electric torpedoes had a reputation for turning in circles, and when this one was only a few yards ahead of *Tang*'s bow it suddenly broke surface and started round in an arc to port. O'Kane speeded up and threw the helm over but the rogue fish hit abreast of *Tang*'s after torpedo room with a shattering explosion. O'Kane and the other eight men on the conning tower were swept into the sea. Three, including O'Kane, lasted through the night and were picked up by a Japanese escort vessel, where they were clubbed and beaten by survivors of the ships they had sunk. Lieutenant Larry Savadkin escaped from the flooding conning tower when the sinking *Tang* reached fifty feet.

Men in the control room managed to close the conning tower hatch, though it leaked, and levelled off the boat by manual control of the vents. Thirty survivors gathered in the forward torpedo room. Threatened by an electric fire in the forward battery, they held on until 6 a.m.; when the Japanese patrol seemed to have gone, then thirteen men struggled to the surface 180 feet above them, using Momsen Lungs, similar to the British Davis Escape Apparatus, through an escape trunk. Eight gathered on the surface and clung to a buoy. Three floated off and drowned, the others were taken prisoner.

By the end of January 1945 the Allies controlled the Philippines, left isolated Japanese forces there to rot, and moved on to the capture of Iwo Jima in the Volcano Islands and Okinawa in the northern Ryuku chain, from both of which medium bombers could raid Japan, and fighter support could be given to the B29s which had been bombing Japanese cities and war industries since November from Saipan. The Japanese in the East Indies were cut off, and Japan herself was feeling the effects of strangulation by sea power, which had brought blockade literally home to her. American naval might in the Pacific, which included 250 submarines, was augmented by a strong British Pacific Fleet, with carriers, battleship, cruisers and supporting warships.

Jim Fife, who had once observed the Mediterranean war for the US Navy in HMS *Parthian*, was now Commander, Submarines, South-West Pacific, and as such had British submarines under his orders working in the East Indies. The only major Japanese warships left in this area were four heavy cruisers based on Singapore, two of them already damaged by US submarines. Two British submarines, Bulkeley's HMS *Statesman* and Andrew's HMS *Subtle,* were patrolling the Malacca Straits in early May when they sighted *Haguro*, making for Burma with supplies for the Japanese army. Their report brought up ships from Ceylon. *Haguro* was bombed and damaged by a plane from the British escort carrier *Shah* on 16th May and finished off by four British destroyers that night.

Commander A.R. Hezlet, of Mediterranean fame, was patrolling the Java Sea in the submarine HMS *Trenchant* in early June when he was ordered to move to the Malayan coast. On his way there he intercepted contact reports from the US submarines *Blueback* and *Chub* reporting that the *Ashigara* had put into Batavia. Hezlet requested and was given permission to head for the Bangka Strait, though which the cruiser would pass when she returned to Singapore.

Trenchant, having negotiated the minefield in the northern entrance to Bangka, took the inside of the Strait, HMS *Stygian* the other. On 8th June *Blueback* reported that *Ashigara* had left Batavia northbound.

A destroyer came dashing up the Strait, fired at *Trenchant*, and was missed by *Stygian*'s torpedo. Depth-charges were exploding round *Trenchant* as Hezlet sighted *Ashigara* coming north, hugging the Sumatran coast. He fired eight torpedoes from his bow tubes from a range of just over two and a half miles. *Ashigara* saw them coming but was so close to the shore that she had only limited movement, to seaward. Five hits reduced her to a wreck, though she moved ahead sufficiently to avoid Hezlet's stern torpedoes, and opened fire briefly on *Trenchant*'s periscope. Her destroyer escort returned in time to pick up survivors.

Soon after this one of the British four-man 'midget' submarines of the type which had disabled the great battleship *Tirpitz* in a Norwegian fiord, penetrated the Singapore harbour defences and finished off the *Takao*, hit by a torpedo from McClintock's *Darter* on 23rd October, with an explosive charge.

The taking of Okinawa drained Japanese traffic out of the East China Sea. Now the only safe approach to the Asian mainland for Japanese ships lay across the Sea of Japan, where *Wahoo* had been lost in 1943. Since then mines and patrols, adding to the viciousness of the currents, had made it no-go for American submarines. The northerly entrance through the La Pérouse Strait, which Morton had used, and Tsugaru Strait were both virtually impassable, but there was just a possibility that the Tsushima Strait to the south might be navigable, with the use of a new FM sonar which could detect a mine at 700 yards, though a good operator was needed to distinguish the ringing tone of 'Hell's Bells' from fish, kelp or thermal layer echoes.

Armed with FM sonar, the nine submarines of a wolf pack

christened Hydeman's Hellcats, from the name of its commander, attempted the passage of the Tsushima Strait.

The Hellcats left Guam on 27th May, led by Hydeman in *Sea Dog*. En route, *Tinosa* rescued ten men from a crashed B29. On being told where the boat was going one of the fliers said, 'Hell, Commander, throw us back!' They were transferred to *Scabberdfish*, which was not going through the mines of Tsushima.

They reached the entrance on 3rd June and split up into groups for the passage through. That night *Sea Dog* led Steinmetz's *Crevalle* and Germerhausen's *Spadefish* into the western channel between Tsushima Island and the southern coast of Korea.

Eighteen fraught hours later they surfaced in the Sea of Japan, and the next group, Pierce's *Tunny*, Lynch's *Skate* and Edge's *Bonefish*, prepared to go that night. *Skate* was scraped by a mine cable, but they too got through. Risser's *Flying Fish*, Tyree's *Bowfin* and Lathan's *Tinosa* came through on the following night.

The nine Hellcats lay low until 9th June and then burst upon the unsuspecting Japanese shipping. In seventeen days they sank twenty-seven *marus*, and *Skate* torpedoed the submarine *I22*. By 24th June, torpedoes spent, they had all come safely out again escept *Bonefish*, sunk on the 18th by patrol boats.

More boats followed them in July, though only eight *marus* and two escorts could be found to sink there in two months. The war ended with five submarines inside the area. Sennett's *Chub* sank four small *marus* in late July.

On 5th August Lockwood sent a signal ordering all his submarines to pull back to at least a hundred miles from the coastline of Kyushu, and a big carrier strike on Kyushu scheduled for the 7th was cancelled. On 6th August the first atomic bomb was dropped on Hiroshima, and three days later another fell on Nagasaki. On 14th August *Torsk* sank two very unfortunate coast defence vessels there, the last Japanese ships to be sunk by United States submarines.

By that time the 288th US submarine of the war had been commissioned. Over 250 had served in the eastern theatres of war, and at least 185 had each sunk one or more, sometimes many more, enemy ships, amounting to a total of over 1,100 merchantmen and more than two hundred warships destroyed, reducing the pre-war merchant total of some 8,000,000 tons to well under 2,500,000 tons. Japanese admirals asserted after the war that the American submarine, more than any other weapon, had brought about the fall of the Empire, even allowing for the great destruction wrought by carrier planes in their thousands. Of fifty-two US submarines lost, thirty-seven were sunk with all hands. A total of 168 men surviving seven boats sunk by enemy action were repatriated, most of them in a very poor state of health, from Japanese prisoner of war camps.

On 14th May 1954, nine years after the war had ended, *U505*, the submarine captured by Captain Dan Gallery's Task unit on 4th June 1944, off the coast of Africa, began a voyage from the berth where she had been rusting away in Portsmouth, New Hampshire, USA, to Chicago, on Lake Michigan. Threading the St. Lawrence River, the Welland Canal and Lakes Ontario, Erie, Huron and Michigan, she arrived off the Chicago waterfront in June, and on the night of 3rd September was beached and dragged across a main road to a berth in a concrete cradle outside the Museum of Science and Industry, where the American people, whose desperate need had created the first operational submarine boat 178 years before, could see a prime example of the 'weapon of the weaker nation' which in the years of its maturity had brought one empire down and almost destroyed another.

Acknowledgements

I am especially grateful to: the staffs of the Historical Branch (Navy) at the Ministry of Defence, London, the Public Relations Office, Kew, and the Naval Historical Centre, Washington, DC, for their kind help with the locating and interpretation of official records; Rear Admiral Richard O'Kane, US Navy (Ret.) for talking to me about his illustrious USS *Tang,* his service in the USS *Wahoo* and *Argonaut,* and the US Navy's submarine war in the Pacific; and to David Brown, Mike Wilson and John Winton for their kind loan of photographs.

For further information on the U-boat war in World War 1 I am particularly indebted to the book *The Most Formidable Thing* by Rear Admiral Sir William Jameson KBE CB (Hart-Davis, 1965) and, for events leading up to the loss of the *Lusitania,* to *Lusitania* by Colin Simpson (Longman's 1972). For submarines in World War 2, on U-boats Wolfgang Frank's *The Sea Wolves* (Weidenfeld & Nicolson 1955), on British submarines in the Mediterranean David A. Thomas's *Submarine Victory* (William Kimber, 1961) and on US Navy submarines in the Pacific W.J. Holmes' *Undersea Victory* (Doubleday, New York 1966) were particularly helpful sources of information.

K.P.

Bibliography

Admiralty Trade Division: *The Economic Blockade*, 1920.

C.S. Alden and A. Westcott: *The United States Navy*, Robert Hale.

Charles Anscombe: *Submariner*, William Kimber 1957.

Richard Baxter: *Stand By To Surface*, Cassell 1944.

Commander Edward Beach: *Submarine!*, Heinemann 1953.

Geoffrey Bennett: *Naval Battles of World War 1*, Batsford 1968.

Jochen Brennecke, trans. R.H. Stevens: *The Hunters And The Hunted*. Burke Publishing Co 1958.

Rear Admiral Gordon Campbell VC, RN: *My Mystery Ships*, Hodder & Stoughton 1928.

Robert Carse: *Blockade*, Rinehart & Co. Inc, New York 1958.

John Costello and Terry Hughes: *The Battle Of The Atlantic*, Collins 1977.

Admiral Karl Dönitz, trans. r.H. Stevens: *Memoirs. Ten Years And Twenty Days*. Weidenfeld & Nicholson 1958.

Wolfgang Frank: *The Sea Wolves*. Weidenfeld & Nicholson 1955.

Rear-Admiral Daniel V. Gallery, USN (Ret): *We Captured A U-boat*, Sidgwick and Jackson 1957.

R.H. Gibson and M. Prendergast: *The German Submarine War*, Constable 1931.

Robert M. Grant: *U-boats Destroyed*, Putnam 1964.

Sydney Hart: *Submarine Upholder*, Oldbourne Book Co., Ltd. 1960.

Ernst Hashagen: *The Log Of A U-boat Commander*, Putnam 1931.

Bodo Herzog: *U-boats In Action, 1939-45*, Ian Allan 1970.

A.A. Hoehling: *The Great War At Sea, 1914-1918*, Arthur Barker 1965.

W.J. Holmes: *Undersea Victory*, Doubleday, New York 1966.

Rear-Admiral Sir William Jameson KBE CB: *The Most Formidable Thing*, Hart-Davis 1965.

Geoffrey Jones: *The Month Of The Lost U-boats*, William Kimber 1977.

Lieutenant-Commander P.K. Kemp: FSA. FR Hist S, RN (Ret.) *Victory At Sea 1939-1945*. Frederick Muller 1952.

Fleet Admiral Ernest J. King, USN: *US Navy At War 1941-1945*, US Navy 1946. (Report to The Secretary of The Navy).

Commander David D. Lewis, USN: *The Fight For The Sea*. New World Publishing Corporation, New York 1961.

Eric Leyland: *Crash Dive*, Edmund Ward 1961.

Peter Liddle: *Men Of Gallipoli*, Allan Lane.

Commander F.A. Lipscombe, OBE, RN: *Historic Submarines*, Hugh Evelyn 1970.

Vice-Admiral Charles A. Lockwood, USN (Ret) and Hans Christian Adamson, Col USAF (Ret.) *Battles Of The Phillipine Sea*, Thomas Y. Crowell Co., New York 1967.

Walter Lord: *Incredible Victory*, Hamish Hamilton 1968.

Paul Lund and Harry Ludlam: *Night Of The U-boats*, Foulsham 1973.

Captain Donald MacIntyre, DSO and two bars, DSC RN (Ret.): *U-boat Killer*, Weidenfeld and Nicholson 1956. *The Battle Of The Atlantic*, Batsford 1961. *Fighting Under The Sea*, Evans Bros., 1965.

Alastair Mars: *Unbroken*, Frederick Muller 1953. *British Submarines At War*, William Kimber.

David Masters: *I.D. New Tales Of The Submarine War* and *Up Periscope*, Eyre and Spottiswood 1975.

James M. Merrill: *The Rebel Shore*, Little, Brown & Co., Boston 1957.

Henry Newbolt: *Submarine and Anti-Submarine*, Longmans Green 1918.

Rear-Admiral Richard O'Kane, USN (Ret.): *Clear The Bridge!* MacDonald & Jane's 1977.

Len Ortzen: *Stories of Famous Submarines*, Arthur Barker 1973.

Peter Padfield: *The Great Naval Race*, Hart Davies.

Roger Parkinson: *The American Revolution*, Wayland.

E.B. Potter and Fleet Admiral Chester W. Nimitz USN: *The Great Sea War*, Harrap 1960.

Fletcher Pratt: *The Compact History Of The United States Navy*, Hawthorn Books Inc, New York 1957.

Anthony Preston: *U-Boats*, Arms and Armour Press 1978.

Grand Admiral Raeder: *Struggle For The Sea*, William Kimber 1959.

Hugh F. Rankin (Editor): *The American Revolution*, Secker & Warburg 1964.

Jurgen Rohwer: *The Critical Convoy Battles Of March 1943*, Ian Allan 1977.

Vice-Admiral Friedrich Ruge, trans. Commander M.G. Saunders, RN: *Sea Warfare 1939-1945*, Cassell 1957.

Commander C.W. Rush, USN, W.C. Chambliss and H.J. Gimpel: *The Complete Book of Submarines*, World Publishing Co, Cleveland and New York 1958.

J.P. Mallmann Showell: *U-boats Under The Swastika*, Ian Allan 1973.

Colin Simpson: *Lusitania*, Longmans 1972.

David A. Thomas: *Submarine Victory*, William Kimber 1961.

Lowell Thomas: *Raiders Of The Deep*, Doubleday, Doran &, Co, New York 1929.

J. F. Turner: *Periscope Patrol*, Harrap 1957.

US Navy Department: *Dictionary Of American Naval Fighting Ships*, 1959.

US Naval History Division: *US Submarine Losses*, Washington 1963.

C.E.T. Warren and James Benson: *The Admiralty Regrets*, Harrap 1958.

Herbert J. Werner: *Iron Coffins*, Arthur Barker 1969.

Arch Whitehouse: *Subs And Submariners*, Muller 1963.

John Winton: *The Victoria Cross At Sea*, Michael Joseph.

Index